THE UNIVERSE

KARL TAYLOR COMPTON *1887–1954*

The Karl Taylor Compton Lectures honor the memory of the ninth president of the Massachusetts Institute of Technology by bringing to the M.I.T. community some of the great minds of our time who contribute to the integration of scientific, cultural, and philosophical concerns — a synthesis richly achieved by Karl Taylor Compton and shared with colleagues and students during his long leadership of the Institute.

KARL TAYLOR COMPTON LECTURERS

Niels Bohr *1957*

Otto Struve *1959*

André Lwoff *1960*

Isidor I. Rabi *1962*

KARL TAYLOR COMPTON LECTURES

THE
UNIVERSE

by

Otto Struve

The M·I·T Press

Massachusetts Institute of Technology

Cambridge

Massachusetts

PREFACE

The Karl Taylor Compton Lectures in Astronomy were presented at the Massachusetts Institute of Technology in November, 1959. Although I had not known Karl Compton intimately, I had met him at various meetings. I remember especially a large meeting of the American Association for the Advancement of Science in Minneapolis at which he presided when I gave an account of progress in astrophysics during the first thirty-five years of this century. Compton encouraged me to continue with my own research, and on subsequent occasions he often helped me with his wise counsel. I was, of course, much more closely associated with his brother, Arthur Compton, who served as professor of physics at the University of Chicago during most of the years I spent at the Yerkes Observatory. In the last few years of my directorship at the Yerkes Observatory, Arthur Compton was my immediate superior as Dean of the Division of Physical Sciences, of which the Observatory was a part.

In presenting these lectures, I have chosen a few topics that seemed of interest to me, and I have not made an attempt to cover the whole large field of astronomy. Hence, this book should be regarded as a selection of topics rather than as a complete account of astronomical progress.

I am especially indebted to the Compton Lecture Committee and to its chairman, Professor F. O. Schmitt of the Massachusetts Institute of Technology, who made all the arrangements for me and who made my stay in Cambridge pleasant and interesting.

OTTO STRUVE

Green Bank, West Virginia
January, 1961

CONTENTS

I. THE SOLAR SYSTEM: ITS ORIGIN AND EVOLUTION

Is the solar system unique in our galaxy? This is the basic question that confronts the modern astronomer when he attempts to discuss the problem of the origin and evolution of its central star, the sun, its nine attendant large planets and their satellites, its millions or even billions of small solid bodies, called minor planets (or asteroids) when they have diameters of the order of 1 to 500 miles, and meteors when they are a few millimeters in size, its 100 billion comets, its dust which can be seen as the zodiacal light and the counterglow on a clear, moonless night, and its hydrogen gas which remained undetected until powerful spectrographs mounted in high-flying rockets by the scientists of the Naval Research Laboratory detected it a few years ago.

Of all the many problems of astronomy, the most fascinating by far is the study of the origin and evolution of the solar system (Figures 1 to 3). Undoubtedly every astronomer, no matter what his special work may be, has thought about it and weighed in his mind the advantages and disadvantages of the different hypotheses that have been advanced during the past two hundred years; more than that, he has probably been hoping to contribute, directly or indirectly, to its solution.

As astronomers, we often pride ourselves with being the disciples of the oldest among the sciences. This has its advantages, but it also brings with it a tendency to become steeped in tradition. When Immanuel Kant, in 1755, laid the foundations for all later cosmogonical speculations, a great deal was already known about the prop-

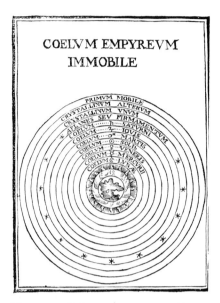

Figure 1. Solar system according to Aristotle.

(Illustration by Peter Gassendi, *Institutio Astronomica*, London, 1653, page **4**.)

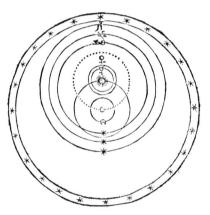

Figure 2. Solar system according to Tycho Brahe.

(Illustration by Peter Gassendi, *Institutio Astronomica*, London, 1653, page 168.)

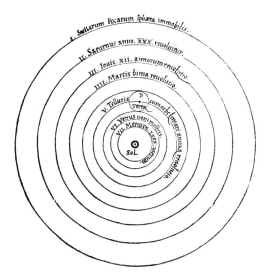

Figure 3. Solar system according to Copernicus.

(Illustration by Nicolas Copernicus, *De Revolutionibus Orbium Coelestium*, Nurenberg, 1543, page 10.)

erties of the solar system but hardly anything about the fixed stars. It was, therefore, natural to concentrate upon trying to explain the regularities that had been observed among the orbits of the planets, their axial rotations, satellites, and so on. Many of these regularities cannot be due to chance; for example, the fact that all the planets move in approximately the same plane and in the same direction must be a consequence of the manner in which they were formed. Most later workers continued Kant's efforts to explain these regularities by inventing a suitable primordial medium, such as the hot, spinning, and contracting nebula of Laplace, out of which the sun and the planets were formed, and endowing it with such physical properties as were needed to explain the dimensions of the planetary orbits, the relative masses of the sun and the nine large planets, and so forth (Figures 4 to 10); or they postulated an appropriate event, such as the close passage of another star in the theories of Moulton and Chamberlin, of Jeans, and of Jeffreys. The hypothetical media and their properties, and the catastrophic events were adjusted in

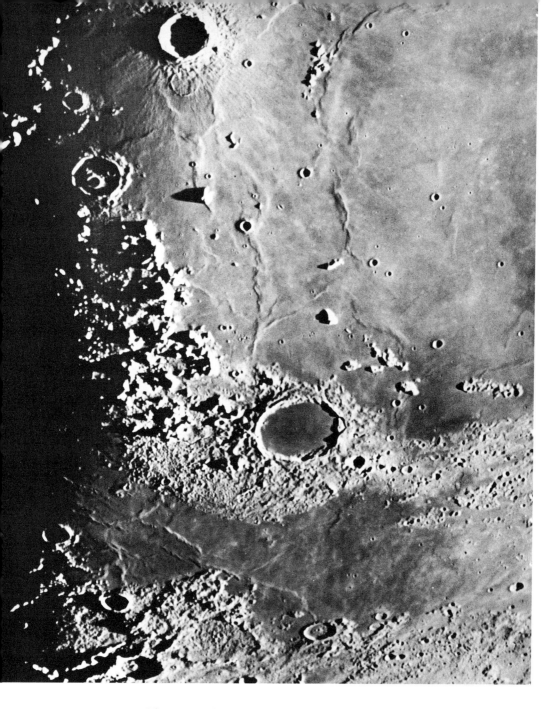

Figure 4. Photograph of the lunar surface. Crater with the flat bottom, just below the center of the photograph, is Plato. The mountains to the left of it are the Alps. The sharp peak is Piton, and the crater at the top is Archimedes. On the right is Mare Imbrium.

(Yerkes Observatory photograph.)

Figure 5. Photograph of the lunar surface. The crater in the middle of the photograph is Alphonsus; above it is Arzachel, and below it is the large crater Ptolemaeus. On the upper right, the "straight-wall" is located in Mare Nubium. *(Yerkes Observatory photograph.)*

Blue

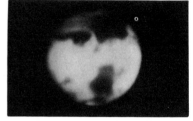

Red

Figure 6. Venus, in blue light; crescent phase.

(*Mount Wilson and Palomar Observatories photograph with the 200-inch telescope.*)

Figure 7. Mars, in blue and red light.

(*Mount Wilson and Palomar Observatories photograph with the 200-inch telescope.*)

Figure 8. The planet Jupiter, in blue light, showing large red spot, satellite Ganymede, and shadow above.

(*Mount Wilson and Palomar Observatories photograph with the 200-inch telescope.*)

Figure 9. Saturn and ring system.

(Mount Wilson and Palomar Observatories photograph with the 100-inch telescope.)

Figure 10. Saturn in blue light.

(Mount Wilson and Palomar Observatories photograph with the 200-inch telescope.)

such a way as to permit the authors of the various hypotheses to deduce many, if not all, of the observed regularities of the solar system.

In the earlier stages of cosmogony this was a reasonable procedure. No one knew whether formations of the kind Laplace and others had proposed really existed in the galaxy, or whether events of the type considered by Moulton and Chamberlin actually take place among the stars. One could always take refuge in the thought that such formations and events *could* have existed; and even if future observations should ultimately fail to produce them, there was always the comforting excuse that for all we know the solar system might be unique in our galaxy.

But gradually astronomers began to study the fixed stars. In 1837 and 1838 the first stellar distances were determined, and fifty years later the "local swimming hole" of our galaxy (to use an expression coined by Walter Baade for describing a region of space surrounding the sun to a distance of about 300 light years) had been fairly well explored. Then, in 1913, came H. N. Russell's great advance, which culminated in the construction of what is now called the "Hertzsprung–Russell diagram" — a relation between the intrinsic luminosities of the stars and their radii (or surface temperatures) (Figure 11).

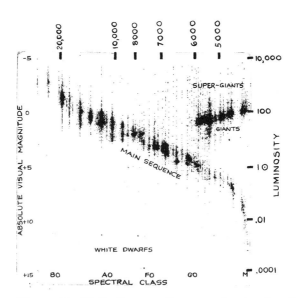

Figure 11. H–R diagram. Spectrum-luminosity diagram. The points represent 6,700 stars.

(Diagram by W. Gyllenberg, Lund Observatory.)

The more we learned about the stars, the more we realized that the sun is a fairly normal star of what we describe as the "main sequence" in the H–R diagram — a relatively cool dwarf having about the usual three principal physical parameters: a mass of 2×10^{33} grams, a radius of 7×10^{10} centimeters, and an energy output of 4×10^{33} ergs/second.

If we imagine ourselves on our nearest stellar neighbor, α Centauri, about 4 light years, or 4×10^{18} centimeters, away, the sun would appear as a fairly bright star, perhaps like the well-known star Capella, but the planet Jupiter, about 4 seconds of arc away, would be completely lost in the glare of the brilliant sun and would not be recognizable except on long photographic exposures with the largest existing telescopes if the glare of the sun could somehow be eliminated, which, of course, cannot now be accomplished.*

It is, therefore, not at all surprising that no one has ever seen or photographed a planet belonging to any other star than the sun: the limitations of our instruments account for this. There are perhaps other methods that may in the not very distant future enable us to discover such planets — by observing the decrease in the apparent brightness of a star when a planet like Jupiter accidentally transits across its luminous disk and eclipses 1 per cent of the latter; or by observing the small orbital motion of the star by means of the Doppler effect in the period of the planet. It has also been suggested by N. Roman that a large telescope equipped with coronograph techniques and mounted in a space satellite might be capable of recording directly the faint light of a distant planet resembling Jupiter, if such a planet does exist in the system of α Centauri. But as yet the evidence is wholly negative.

Nevertheless, our hypothetical observer on α Centauri would be able to establish that the sun has a surface temperature of about 6,000°K, and that its Fraunhofer absorption lines are exceedingly sharp and narrow. If his spectrographic techniques resemble ours,

---

* J. C. Pecker has recently published an amusing story under the title *Les systèmes planétaires dans l'univers*, Société Astronomique de France (1960), in which he describes what "Dr. C. N. Blackbody" of the Sipar Observatory on planet Erreth of the system of α Centauri observed when he pointed his telescope at the star χ-1 Cassiopeiae — which we call the sun.

he would conclude that the sun rotates very slowly around its axis; otherwise, its absorption lines would appear conspicuously blurred. He might even succeed in detecting the existence of temporary dark areas on the sun — the sunspots — and determine from their periodic appearances and disappearances that the rotation period of the sun is about twenty-five days and, consequently, its rotational velocity at the equator is only 2 km/sec. This result would not disturb him: he would know, as we do, that all the rest of the cool dwarfs in the galaxy which are not members of close double-star systems have slow axial rotations. He would also know that many other stars — much hotter than the sun and several times greater in size — usually have equatorial rotational velocities of the order of 100 km/sec, and in exceptional cases even 500 km/sec. Their rotational periods are only a few hours in duration.

If the hypothetical observer knew anything at all about the cosmogonical theories of our predecessors, he would probably immediately discard all those that attribute the origin of the planets to a rapidly spinning star of the solar type: there are not any such stars among the billions of cool dwarfs in the galaxy.

He might, of course, be tempted to believe that the old Laplacian ideas could be applied to those nonsolar type stars that have rapid rotations. Those whose equatorial rotational velocities are of the order of several hundred kilometers per second can be shown to be unstable at their equators: the centrifugal force exceeds the force of gravity, and observations show that they do, in fact, shed gas in a narrow disk-like nebula whose existence is revealed by the presence of double emission lines of hydrogen, and sometimes other atoms, from the approaching and receding limbs of the nebulous disk. But the density of these formations is exceedingly small — at any given time the total mass of the revolving disk is of the order of 10^{-5} or 10^{-6} of the mass of the star. By contrast, the mass of Jupiter alone is about $\frac{1}{1000}$ of the mass of the sun; apparently, therefore, no planets can be formed out of these tenuous nebulosities.

We see that the study of the stars must have a bearing upon the development of cosmogonical thought. Since all those properties of the sun which we observe in other stars are normal for a very large group of reddish dwarfs in the galaxy — perhaps as many as 50 billion — it is at least plausible that those other properties that we cannot now test outside the solar system are also common to all or most of them. In other words, it is far more reasonable to start with the working hypothesis that planets are normally present in the vicinity of cool dwarfs than it is to suppose that our own planetary system is unique or very rare. But if we make this assumption, we are immediately led to consider the evolution of the sun and its system of planets in the light of what we know, or surmise, concerning the evolution of the stars in general. We should satisfy, as nearly as is possible, not only the observed regularities in the solar system but also the observed properties of stars and nebulae in various stages of their evolution.

This procedure has required breaking with the old tradition and departing entirely from the conventional form of cosmogonical research. The transition was not an easy one, and it required a great deal of time. But gradually astronomers in several countries made the break: in Germany, C. F. von Weizsäcker adapted the theory of turbulence to ordinary stars and tried to explain why the hot stars have, on the average, large rotational velocities, while the cool stars have very slow rotations. Another step in the same direction was made in my Vanuxem lectures at Princeton University in 1949. In Russia, V. A. Ambartsumian has stressed the departures from traditional cosmogony, and V. G. Fessenkov, partly in collaboration with A. G. Masevich, has discussed the possible effects of loss of mass by stars in the form of "corpuscular radiation." In America, the most significant contributions were made by G. P. Kuiper at the Yerkes Observatory.

Our task, then, is to try to explain simultaneously the observed properties of the solar system and of the billions of solar-type stars in the galaxy. We shall start with the former. What is the size of the solar system? Until recently we assumed that the most distant planet, Pluto, marked the outer edge of the solar system. If we use the distance between the sun and the earth as our unit — the astronomical unit is 93 million miles, or 1.5×10^{13} centimeters (the radius of the earth is 6,400 kilometers or 6.4×10^8 centimeters) — then the distance from the sun to Pluto is about 40 astronomical units, or about $\frac{1}{7000}$ of the distance to the nearest star. But recent

Figure 12. Comet Brooks, photographed on October 19, 1911.
(*Yerkes Observatory photograph.*)

work, especially by J. H. Oort of Leiden, has shown that while no large planetary bodies are now known beyond the orbit of Pluto, there are some 10^{11} comets (one of which is shown in Figure 12), all belonging to the sun's family, which travel in extremely elongated orbits around the sun and reach out to distances of the order of 150,000 astronomical units — more than one-half of the distance from us to the star α Centauri. We must, therefore, assign to the solar system a correspondingly large volume of space. It seems as though we can, in effect, assume that interstellar space is subdivided into a large number of adjoining cells, one for each star, which in the "local swimming hole" of the galaxy can be represented by cubes whose sides measure roughly 4 light years (Figure 13). One light year is the distance traveled by a ray of light in 1 year, or 3×10^7 seconds:

$$1 \text{ light year} = 3 \times 10^{10} \times 3 \times 10^7 \approx 10^{18} \text{ centimeters.}$$

The volume of each cell is thus:

$$\text{Volume of cell} = (4 \times 10^{18})^3 \approx 10^{56} \text{ cm}^3.$$

The mass of the solar system is almost wholly concentrated within the sun (Figures 14, 15): all the planets, satellites, and the like, contribute only about 0.1 of 1 per cent. The mass of the sun, easily derived from Kepler's third law, is 2×10^{33} grams. If we assume

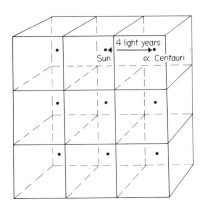

Figure 13. Schematic representation of the cells of interstellar space dominated by individual stars.

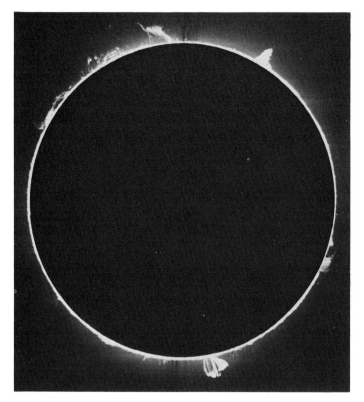

Figure 14. Solar prominences. Whole edge of sun taken with calcium K line, December 9, 1929.

(*Mount Wilson and Palomar Observatories photograph.*)

that the solar system has condensed out of some primordial diffuse gas that originally filled the cell uniformly, we could derive its original density by spreading out the present mass of the solar system over the volume of the cell:

$$\delta = \text{density of medium} = \frac{M}{V} = \frac{2 \times 10^{33}}{10^{56}} = 2 \times 10^{-23} \text{ gm/cm}^3.$$

Since the average atom of the original gas was mostly hydrogen, we may assume that the mass of one particle

$$\mu = 2 \times 10^{-24} \text{ gram.}$$

Figure 15. Solar corona. Photographed during eclipse of the sun on September 22, 1922.

(*Mount Wilson and Palomar Observatories photograph.*)

Hence, the number of atoms in the medium must have been of the order of

$$N = \frac{2 \times 10^{-23}}{2 \times 10^{-24}} = 10 \text{ atoms/cm}^3.$$

Is this result reasonable? If it is, it would give us greater confidence in the theory of Kant, which attributes the origin of the solar system to a process of gravitational contraction within a primordial gaseous medium of approximately uniform density. The answer is decidedly yes: we can measure the number of interstellar atoms in

1 cubic centimeter from the intensities of the insterstellar absorption lines produced along the entire path between us and some distant star. For the "local swimming hole" this turns out to be about 1 atom per cubic centimeter, or somewhat less. Evidently the stars have absorbed within their masses some 90 per cent of the original gas, leaving about 10 per cent in the form of diffuse interstellar gas.

Incidentally, the observed density of the interstellar medium, about 1 H atom/cm³ is the *average* density for our region of the galaxy. In certain regions, for example, the Orion nebula (Figure 16), the density of the gas may be as great as 10^3 or even 10^4 atoms per cm³. But these regions are relatively scarce in the Milky Way.

We must now investigate whether an original medium having a density of the computed order of magnitude, about 10 atoms/cm³, would have been capable of condensing into a star. To do this we shall first consider the problem of tidal instability of a condensation. Suppose that as a result of turbulence an eddy in the medium is formed, whose density is slightly greater but still very close to the average value of 10 atoms/cm³. Will it retain its identity and grow by gravitational attraction and accumulate within its body other atoms or eddies? The mass of the galaxy as a whole is concentrated mostly in its inner regions. There are, of course, many stars and nebulae at distances from the galactic center that exceed the distance of the solar system from this center, and in an accurate calculation their effect must be allowed for. But for a crude, order-of-magnitude estimate, we may assume that the concentration that led to the formation of the solar system was located at a distance of about 30,000 light years from the galactic center, and that at this center there is a mass of about 2×10^{11} solar-type stars, or about $2 \times 10^{11} \times 2 \times 10^{33} = 4 \times 10^{44}$ grams. We consider the disrupting force of the tides produced by the central mass upon the newly formed condensation (Figure 17).

Assume that the condensation can be regarded as two spherical bodies, each of mass m, whose centers are separated by $2r$. Then the attraction of these two bodies upon each other is

$$G \frac{m \times m}{(2r)^2} = G \frac{m^2}{4r^2}.$$

Figure 16. The great nebula in Orion.

(*Yerkes Observatory photograph.*)

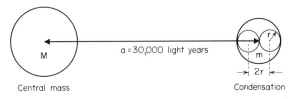

Figure 17. Tidal instability, as applied to the formation of planets.

The gravitational attraction of the center upon the nearer body is $G[Mm/(a-r)^2]$ and upon the more distant body $G[Mm/(a+r)^2]$. The difference,

$$GMm\left[\frac{1}{(a-r)^2} - \frac{1}{(a+r)^2}\right] \approx GMm\frac{4r}{a^3},$$

is the tidal disrupting force. The condensation will be stable if the mutual attraction of the two bodies exceeds the disrupting force:

$$G\frac{m^2}{4r^2} > GMm\frac{4r}{a^3},$$

or

$$\frac{m}{r^3} > \frac{16M}{a^3}.$$

The mass of each body is

$$m = \tfrac{4}{3}\pi r^3\rho,$$

where ρ is its density. Hence,

$$\frac{4\pi r^3\rho}{3r^3} > \frac{16M}{a^3},$$

or, approximately,

$$\rho > \frac{4M}{a^3}.$$

If other forces are present which tend to impede concentration — centrifugal force due to rotation, magnetic force, and so forth, then the numerical coefficient would be greater than 4, and some astronomers, notably Fessenkov, have rather arbitrarily used the expression

$$\rho > \frac{10M}{a^3},$$

and we shall follow them in this calculation. We now substitute the numerical values:

$$\rho > \frac{10 \times 4 \times 10^{44}}{(3 \times 10^{22})^3} \approx 10^{-22} \text{ gm/cm}^3$$

or about 50 H atoms/cm^3. The agreement with the computed average density of the original medium is sufficient to inspire confidence in the conclusion that tidal forces from the Milky Way as a whole would not disrupt the newly formed condensation.

But would the condensation also resist the tidal disrupting force of a nearby star, say, one at a distance of 4 light years, or 4×10^{18} centimeters? We use the same expression, but now

$$M = 2 \times 10^{33} \text{ grams},$$

$$a = 4 \times 10^{18} \text{ centimeters}.$$

In this case

$$\rho > \frac{10 \times 2 \times 10^{33}}{(4 \times 10^{18})^3} \approx 3 \times 10^{-22} \text{ gm/cm}^3.$$

This result is somewhat larger than the one obtained previously. I believe we must conclude that the medium in the original cell already had some inhomogeneities. This is consistent with the observations of nebulae and interstellar absorption lines which show that the present medium is highly inhomogeneous and consists of regions of greater and smaller average densities, as can readily be inferred from the photographs.

We can also conclude that the disrupting effects of stars much closer than 4 light years, for example, in dense star clusters, imply that the original medium out of which the members of clusters were condensed must have been by several orders of magnitude more dense than the average. The Orion nebula, for example, with its average density of 10^3 to 10^4 H atoms/cm^3 could have given rise to the formation of a star cluster; and observations show that such a cluster does, in fact, exist in the central parts of the nebula.

The tidal stability criterion represents a necessary, but by no means a sufficient, criterion for the formation of a star. A hot gas will expand and not contract, unless the gravitating mass is very great. This argument has, in fact, been used by L. Spitzer, Jr., to demolish all those theories of the origin of planets which attribute them to the condensation of a very hot and large prominence erupted from the sun by the attraction of a passing star. Consider this aspect of the problem slightly more in detail but in a simplified form suggested by S. S. Huang:

In a sphere of uniform density ρ and of radius R, the total mass is

$$M = \frac{4}{3} \pi R^3 \rho.$$

Let there be a particle on the surface of this sphere whose linear velocity is v. In order that the condensation could retain this particle, its velocity must be smaller than the velocity of escape

$$v < v_{\text{escape}} = \sqrt{\frac{2GM}{R}} = R\sqrt{\frac{8\pi G\rho}{3}}$$

or

$$R > \sqrt{\frac{3}{8\pi G\rho}}\, v.$$

Substituting for the constant of gravitation,

$$G = 6.7 \times 10^{-8},$$

we find

$$R > 1.34 \times 10^3 \frac{v}{\rho^{1/2}}.$$

Take a fairly dense condensation,

$$\rho = 10^{-22} \text{ gm/cm}^3,$$

and assume that the gas is hot, say, $T = 10{,}000°\text{K}$, so that

$$v \approx 10 \text{ km/sec} = 10^6 \text{ cm/sec}.$$

Then

$$R > 1.34 \times 10^3 \, \frac{10^6}{10^{-11}} \approx 10^{20} \text{ centimeters,}$$

or 100 light years. This is much larger than the size of the cell that we had found at the outset — about 2 light years in radius. Only a very cold gas, with $v \sim 0.2$ km/sec, would enable a condensation to grow. Since there is definite evidence that "protostars," in the form of black globules (Figures 18, 19) do occur in connection with hot gaseous nebulae (the temperature of the Orion nebula is about $10^4 \, ^\circ$K), there must be in existence cooling mechanisms such as the formation of dust, or even larger meteor-like bodies, that can radiate heat rapidly and thus meet the requirement of small v.

Although not all the steps we have taken are sufficiently rigorous for a definitive theory, it seems reasonable at this stage to conclude that gravitationally stable condensations of gas can be formed, and

Figure 18. Gaseous nebula NGC 6611 (M 16) in Scutum Sobieski, showing a number of small, round globules and larger, irregular, dark dust clouds.

(*Mount Wilson and Palomar Observatories photograph with the 200-inch telescope.*)

that they grow in mass and are first seen as cold, black globules when they are projected upon the luminous background of an emission nebula. How then should we proceed with the problem of the origin of planets?

Before continuing, I want to dispose of those theories that attribute the formation of planets to the passage of a nearby star. We know from the study of close binary systems, in many of which the surfaces of the two components almost touch one another, that in order to cause spectacular eruptions on one or both stars the near passage must be almost a grazing collision. We might then estimate the probability of such a collision of two stars in the Milky Way. How should we go about this? If we were considering the collision of a blindly walking person with another in the concourse of the Grand Central Station in New York, we might start by giving the distance to our nearest neighbor, say, 20 feet, and our relative velocity, say, 4 feet/sec. Of course all persons are moving about, and in an accurate calculation this should be allowed for. But we shall obtain a surprisingly good order-of-magnitude result if we assume that they remain at rest but that we are moving at the rate of 4 feet/sec. We may then say that no collision can take place until we have traveled 20 feet, which should require $20/4 = 5$ seconds. Let the "cross section" of an average person be 2 feet. To allow for grazing collisions, we allow for our own cross section, also 2 feet. If we move in a random direction, we can reach with equal probability any point on the circumference whose radius is 20 feet. Therefore, in order that there be a collision once in 5 seconds, we divide the sum of the cross sections, 4 feet, by the circumference, $2\pi \times 20 = 126$ feet, and obtain 0.03. After the lapse of the first 5 seconds the probability is, of course, again 0.03 during the next 5 seconds, and so on. We are thus pretty sure that in a few minutes we shall experience at least one collision.

Figure 19. Several small, round, black "globules" are seen projected against the luminous background of the emission nebula NGC 2237 in Monoceros.

(*Mount Wilson and Palomar Observatories photograph with the 48-inch Schmidt telescope.*)

If we apply this reasoning to the stars, and remember that we must consider the problem in three dimensions, we find that with a distance of 4 light years to our stellar neighbor, and a relative velocity of 20 km/sec, it would take us

$$t = \frac{4 \times 10^{13} \text{ km}}{20 \text{ km/sec}} = 2 \times 10^{12} \text{ seconds} = \frac{2 \times 10^{12}}{3 \times 10^{7}} \text{ years,}$$

or about 10^5 years to traverse the required distance.

The radius of the sun, and presumably also of our nearest neighbor, is 7×10^5 kilometers. At a distance of 4 light years, or 4×10^{13} kilometers, this would subtend an angle of about 3×10^{-12} square degree. The target area as seen from the sun would thus be

$$\frac{3 \times 10^{-12}}{4 \times 10^{4}} \approx 10^{-16}$$

of the entire sky, because there are about 40,000 square degrees on a sphere. Our target area, as seen from the other star, would also be 10^{-16} of the sky. Thus the probability of one collision occurring in 10^5 years is only 2×10^{-16}. Even if we allow for greater effective cross sections, say, 10^{-15} of the sky, we could be reasonably certain that the probability of any individual star colliding with another during the entire past life of the galaxy, say, 5×10^9 years, is of the order of only

$$10^{-15} \times \frac{5 \times 10^{9}}{10^{5}} = 5 \times 10^{-11}.$$

Since the number of stars in our galaxy is 2×10^{11}, it is just possible that one or two collisions of the type we are considering have taken place during the past 5 billion years. If the sun was one of the partners of such a collision and the planets resulted from it, the solar system would be unique — and this, as we have already seen, is contradicted by the evidence from other solar-type stars. In deriving this estimate we made the assumption that the stellar motions are distributed at random. This is approximately correct for the vicinity of the solar neighborhood. The fact that the entire group of nearby stars, including the sun, is traveling at a high rate of speed in an approximately circular orbit around the distant galactic center makes no difference.

It should be remembered that even though stellar collisions are exceedingly rare in any one galaxy, the number of distant galaxies observable with the largest telescopes is almost as great as the number of observable stars in our galaxy. Stellar collisions must occur in some of them. What the observable effect would be is not known. Novae, which occur at the rate of several dozen each year in our galaxy, and even supernovae, which occur at intervals of about 300 years, are far too frequent. The chances are that a head-on collision of two stars would produce a splash of incomprehensible magnitude.

Before we leave this topic, it would be well to state that the probability of a passing star "colliding" with the "sphere of action" of a visual binary is by no means negligible. If the semimajor axis of the binary is 50 astronomical units, or $50 \times 1.5 \times 10^8$ kilometers, the target area is roughly 10^8 times larger than in the case of a single star. Hence, if during the lifetime of the galaxy only one or two stars have ever collided with any other star, there may have been as many as 10^8 "collisions" with binary-star orbits. This problem was investigated by R. A. Lyttleton in connection with his theory of the origin of planetary systems. It is improbable that this particular application of the theory can be maintained. The general problem of the perturbations of binary orbits by passing stars has not been investigated, and we do not know what the consequences of such an event are likely to be. It is, however, improbable that such events will help in explaining the principal mystery of binary stars, for example, the frequent occurrence in the Milky Way of very close binaries in which the component stars are almost in contact.

I have already referred to the cosmogonical significance of the distribution of stellar rotational velocities. High rotational velocities occur only in hot stars (spectral types O, B, A, and early F) and never in cool dwarfs (spectral types late F, G, K, and M), in supergiants, Cepheid variables, and long-period variables. Table 1, taken from C. W. Allen's *Astrophysical Quantities* (The Athlone Press, London, 1955, page 186), summarizes the observational results.

Table 1. Stellar Rotation.

Spectrum	v_e (km/sec) Maximum	v_e (km/sec) Mean	$v_e \sin i$ Mean	Distribution of stars in per cent v_e (km/sec) 0	50	100	150	200	250	300	500
Oe, Be	500	350	275	0	0	0	1	3	18	78	
O, B	250	94	73	21	51	20	6	2	0	0	
A	290	112	87	22	24	22	22	9	1	0	
F0 to F2	250	51	40	30	50	15	4	1	0	0	
F5 to F8	70	20	16	80	20	0	0	0	0	0	
G, K, M	5	0	0	100	0	0	0	0	0	0	

Table 1 shows that there is a conspicuous discontinuity at spectral type F5 (surface temperature 6,500°K): the hotter stars tend to rotate rapidly, the cooler ones (which resemble the sun) have very slow rotations. Since the other stellar parameters fail to show an abrupt change at F5, we must infer that some unobservable property accounts for the discontinuity.

What this unobservable property may be is rendered intelligible if we consider what the rotation of the sun would be if there were no planets but all their orbital (and rotational) angular momentum were added to that of the sun. The angular momentum is $m \times v \times r$. Thus for Jupiter the mass is $m = 2 \times 10^{30}$ grams, the orbital velocity is $v = 1.5 \times 10^6$ cm/sec, and the distance from the sun is $r = 7.8 \times 10^{13}$ centimeters. Therefore, its orbital angular momentum is

$$A\ (\math{2\!\!\!|}) = 2 \times 10^{50} \text{ gm cm}^2/\text{sec.}$$

To get a crude estimate for the sun, we may assume that a representative gram has

$$\bar{v} = 0.5 \text{ km/sec} = 5 \times 10^4 \text{ cm/sec,}$$

and a distance from the sun's axis of rotation

$$\bar{r} = 5 \times 10^{10} \text{ centimeters.}$$

Hence,

$$A\ (\odot) = 2 \times 10^{33} \times 5 \times 10^4 \times 5 \times 10^{10}$$

$$= 5 \times 10^{48} \text{ gm cm}^2/\text{sec.}$$

In reality, using a more refined calculation, we can show that the planets carry about 98 per cent of the total angular momentum of the system, the sun the remaining 2 per cent. But the planets carry only $\frac{1}{1000}$ of the mass of the system. If all the present planets were combined with the sun, its equatorial velocity of rotation would be increased by a factor of about 50 and would be close to 100 km/sec; the sun would be a fairly rapidly spinning star and would not, in this respect, differ from the hotter stars.

If, as we shall see, the mass of the original "protoplanets" was much greater than 10^{-3} $\odot$, the sun, as a single star, would have acquired an even faster rotation.

The small angular momentum of the sun and the large angular momentum of the planets have constituted the principal stumbling block of all the earlier cosmogonical theories. Laplace's theory fails to account for this phenomenon as do several of its modifications. Attempts have been made to explain the discrepancy. Alfvén has suggested that the magnetic field of the sun, revolving with the latter, would tend to transfer momentum from the sun to a nebula associated with it. Fessenkov has suggested that the sun may have been, some 5 billion years ago, ten times more massive than now, and that it has lost mass by corpuscular radiation from the surface layers, which has carried off a large part of the original momentum. None of these explanations accounts for the sudden break in v_{equ} at spectral type F5.

On the other hand, we derive confidence in our working hypothesis that the distribution of angular momentum between the sun and the planets was produced by the manner in which the planets were formed, from several recent studies, by Sandage, Slettebak, Herbig, and others. Many yellow giants of types F5 to G have fairly rapid rotations; thus, ξ Geminorum, of type F5 and absolute magnitude $+1.9$, has $v_e \sin i = 73$ km/sec, according to Oke and Greenstein. The rotation of an average F5 dwarf is close to zero. But ξ Geminorum is an old star, which several billion years ago was a dwarf whose radius was only one-half of what it is now, and its spectral type would then have been A (temperature 10,000°K). Because of its smaller size, at the time it was a dwarf its rotational velocity would have been twice as rapid: v_e (original) $=$

$2 \times 73 = 146$ km/sec — a perfectly reasonable value for an A dwarf. Evidently the old stars that increase in size because of the nuclear exhaustion of H in their interiors preserve their original angular momenta; their values of v_e decrease but, because they also become cooler, these smaller values of v_e are still much greater than those observed among the dwarfs of equal temperature.

We conclude that in all probability the angular momentum of a star, with or without planets, remains reasonably constant and is not seriously altered by magnetic braking or other similar processes.

The evidence thus seems to be overwhelmingly in favor of our conclusion that all or most solar-type stars possess planetary systems resembling our own. But this does not mean that we have explained the mechanism by means of which the planets were formed. Nor have we explained how any star, single or with planets, could have acquired as small an angular momentum as that of the entire solar system:

$$A = 3 \times 10^{50} \text{ gm cm}^2/\text{sec.}$$

Observations of nebulae, out of which the stars were formed, show turbulent motions of the order of 5 to 10 km/sec. These are chaotic motions, sometimes in well-defined stream patterns. If such a nebulous mass, containing 2×10^{33} grams, contracts from an initial density of 10^{-22} gm/cm^3 to that of the sun, 1.4 gm/cm^3, these chaotic motions will never completely cancel out: there will remain a residual angular momentum. This is consistent with the principal regularity of the solar system: the orbital planes of the planets and satellites and the equator of the sun coincide within a few degrees. But the angular momentum that is actually observed is much too small.

Suppose that the original cell of the medium out of which the sun was formed had a spherical shape with

$$R = 2 \text{ light years} = 2 \times 10^{18} \text{ centimeters.}$$

Since we have already concluded that the medium was inhomogeneous, suppose that we think of it as composed of N blobs of gas moving about at random with velocities of

$$5 \text{ km/sec} = 5 \times 10^5 \text{ cm/sec.}$$

The mass of each blob would then be

$$M/N = 2 \times 10^{33}/N \text{ grams.}$$

We now apply the statistical theory of "random flights" (following a recent procedure by W. H. McCrea) and compute the expected angular momentum of the mass

$$A = \sqrt{N}\,(M/N)\,V\,R = M\,V\,R/\sqrt{N}$$
$$= 2 \times 10^{33} \times 5 \times 10^5 \times 2 \times 10^{18}/\sqrt{N} = 2 \times 10^{57}/\sqrt{N}.$$

The observations by G. Münch and O. C. Wilson indicate that in the Orion nebula there are regions of greater and smaller velocity than the average, whose angular dimensions are of the order of a few seconds of arc. We may, therefore, assume that N must be greater than 1, but perhaps not greater than 100 or 1,000. Even if $N = 10^4$, the observed angular momentum being 3×10^{50}, the discrepancy corresponds to a factor of 10^5. For the moment we shall disregard this difficulty: it can, I believe, be overcome either by assuming that stars form only in unusually quiescent regions of the interstellar medium or by the same mechanism which McCrea has used in his new theory (presented at the 1959 Liége symposium on stellar evolution) in order to explain the small angular momentum of the sun. We, therefore, assume that the material in the original cell has somehow contracted until it forms a sphere several times greater in radius than 40 astronomical units (the distance of Pluto from the sun). Within this larger sphere we envisage a sphere of radius

$$R = 5 \times 10^{14} \text{ centimeters (about 40 astronomical units)}$$

whose total mass

$$M = 2 \times 10^{33} \text{ grams,}$$

but which contains N blobs whose random directed velocities are 1 km/sec (the gas would be cold, $T = 100°K$). The total angular momentum of this accumulation of N blobs is 3×10^{50} gm cm^2/sec.

We now set

$$M\,V\,R/\sqrt{N} = 3 \times 10^{50} \text{ gm cm}^2/\text{sec, and find } N = 10^5.$$

The mean half distance of two neighboring blobs, r, is obtained from the fact that $Nr^3 = R^3$. Hence, $r = 10^{13}$ centimeters.

McCrea assumes that the mean free path of a blob is about R, for if it were much smaller the blobs would interact. Then if s is the mean radius of a blob, a cylinder of volume $4\pi s^2 R$ would contain one blob and $s = 10^{12}$ centimeters.

The original "protosun" was presumably itself an average blob that had accidentally coalesced with one or two others, and thus began to serve as a center of attraction.

If all the original N blobs, located inside the sphere of radius R, had coalesced with the protosun, the latter's angular momentum would now be 3×10^{50} and not $\frac{1}{50}$ of it. Hence, McCrea suggests that only those blobs combine to form the sun whose original direction of motion is such that they strike the target area of the protosun whose radius must have been somewhat larger than the radius of an average blob, 10^{12} centimeters. McCrea uses a value of approximately 10^{13} centimeters. We thus encounter the problem of collisions all over again. As long as the protosun had a small mass, the motions of the individual blobs were practically rectilinear, and the problem is (1) what fraction of blobs have the required directions to strike the target and (2) what is the average angular momentum of these blobs?

Those blobs that do not hit the target will ultimately move out of the sphere of radius R; but they will be replaced by an equal number of blobs from outside the sphere that have the required directions. Thus the total mass of the sun will ultimately become

$$N\frac{M}{N} = 2 \times 10^{33} \text{ grams.}$$

The average angular momentum of the N blobs that form the sun is, following McCrea, about in the ratio

$$\frac{A \text{ (sun)}}{A \text{ (total)}} = \frac{r}{R} = \frac{10^{13}}{5 \times 10^{14}} = 2 \text{ per cent.}$$

A slightly more elaborate computation by McCrea, allowing for the gravitational action of the growing protosun, leads to 4 per cent, which is still a very reasonable value. The interval of time required for the sun to accumulate N blobs of suitable angular momentum is estimated by McCrea to be about 10^5 years.

McCrea's theory shows that it is possible to account for the small angular momentum of the sun and, by a similar reasoning, for that of most stars. N blobs form the sun, some from within the sphere of radius R, the rest from the outside. The blobs within the sphere of radius R that have large angular momenta supposedly form the planets and their satellites.

But will these blobs, each of mass $M/N \sim 2 \times 10^{28}$ grams actually be capable of coalescing to form a planet like Jupiter, which has a mass of the order of 100 blobs? We have already seen that the criterion of tidal stability can be written in the form

$$\frac{m}{s^3} > \frac{16M}{a^3} \, ,$$

which, with the appropriate numerical values, $m = 2 \times 10^{28}$ grams, $s = 10^{12}$ centimeters, $M = 2 \times 10^{33}$ grams, $a = 5.1$ astronomical units (a.u.) $= 5.1 \times 1.5 \times 10^{13}$ centimeters, and with the arbitrary factor of 2.5 on the right-hand side, results in

$$2 \times 10^{-8} > 3 \times 10^{-7},$$

giving a discrepancy of approximately 10. This is probably not serious: since McCrea's estimate of s is admittedly quite crude and since it appears in the third power in the denominator of the left-hand side, an insignificant reduction in the radii or the blobs would serve to remove the discrepancy.

More serious is the fact, pointed out by McCrea, that two blobs colliding head on, with $v = 1$ km/sec would generate so much heat that the resulting mass would tend to evaporate. It is, however, probable that the relative velocity of two blobs, both moving approximately in the same direction around the protosun is much smaller than 1 km/sec. The collision need not be head on; and there may be in existence dust particles or suitable molecules that are capable of rapidly radiating away a large amount of heat.

The importance of the tidal stability criterion in cosmogony was first clearly demonstrated by G. P. Kuiper, long before McCrea's theory became known. No attempt has as yet been made to integrate the two theories, but it seems to me that they are entirely compatible.

Kuiper assumes that the sun was formed in a fairly dense interstellar cloud, and that a disk-shaped nebula of about 40 a.u. was left behind which revolved around the sun.

The presently observed inclinations of the orbits of the planets give us an indication that this solar nebula had an appreciable thickness at right angles to its plane of original rotational symmetry. The latter may be identified with what we call the "invariable plane" of the solar system. It represents the average for the system, and its orientation in space can be altered only by external forces. If the inclination of the orbit of any one planet changes as a result of perturbations, the inclinations of one or more of the other orbits must change in the opposite direction.

The innermost planet, Mercury, has the largest orbital inclination to the invariable plane (excluding Pluto, which is probably a former satellite of Neptune), amounting to about 6½ ° at a distance from the sun of 0.4 a.u. A protoplanet, moving in such an orbit, indicates, as shown in Figure 20, a thickness of about 0.1 a.u. for the solar nebula. Farther out, say, at Jupiter's distance of 5 a.u. from the sun, the orbital inclination of 0.3 degree again roughly suggests a thickness of 0.1 a.u. for the nebula.

If we assume that the nebula had a cylindrical shape, its volume up to a distance of 30 a.u. from the sun would have been 10^{42} cm³. If we now assume that the original solar nebula contained as much mass as all the planets taken together possess at the present

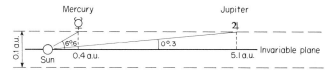

Figure 20. The dashed lines in this diagram represent schematically the limits to the thickness of the solar nebula, if Mercury and Jupiter are included within it at their present inclinations to the invariable plane. The angles are properly labeled, but the vertical scale is exaggerated for clarity.

time, 2×10^{30} grams, the mean density of the solar nebula would have been

$$\rho = \frac{2 \times 10^{30}}{10^{42}} = 2 \times 10^{-12} \text{ gm/cm}^3.$$

Even allowing for possible departures in the assumed values of the thickness and radius of the nebula, it is difficult to see how ρ could have been more than about 10^{-11} gm/cm³.

We apply now the tidal criterion: $\rho > 10M/a^3$ with $M = 2 \times 10^{33}$ grams and, say, $a = 10$ a.u. $= 1.5 \times 10^{14}$ centimeters, to obtain an average result:

$$\rho > \frac{10 \times 2 \times 10^{33}}{(1.5 \times 10^{14})^3} = 4 \times 10^{-9} \text{ gm/cm}^3.$$

The discrepancy is of the order of at least 100: we are forced to conclude that the presently observed masses of the planets account for only 1 per cent, or less, of the mass of the original solar nebula. At least 99 per cent of the nebula must have been driven out of the solar system, and this is consistent with the fact that the terrestrial planets, Mercury ($\mathbf{\yen}$), Venus ($\mathbf{\venus}$), the earth ($\oplus$), and Mars ($\mathbf{\male}$), are strikingly deficient in hydrogen. Even Jupiter ($\mathbf{\jupiter}$) and Saturn ($\mathbf{\saturn}$), which consist mostly of hydrogen, must have accumulated within their bodies only a fraction of the material that was located inside their rings of "dominance." The original mass of the nebula must have been of the order of 0.1 $\odot$, or 2×10^{32} grams. Had all of it been collected into one companion star, the solar system would now be a normal double star of mass ratio $\frac{1}{10}$.

The rather high density of the original solar nebula, about 10^{-9} gm/cm³ is somewhat unexpected. All known galactic nebulae have densities no greater than about 10^{-20} gm/cm³. Neither can the solar nebula be identified with B. J. Bok's dark "globules," which are much larger than 10^{14} centimeters and have much lower densities than 10^{-9}. The question arises whether there are any other objects in the sky today which resemble the sun when it still possessed its nebula.

Seen from the earth at a distance of about 2×10^{15} kilometers (or 200 light years), a nebula with a diameter of 60 a.u. would subtend an angle of about 1 second of arc. Not many stars would be near enough to show such nebulous disks, but spectroscopic effects might be observable. When seen edgewise the path of light through the nebula would be so long that opacity would render the central star invisible. But at right angles to their planes of symmetry the spectrum of the star might show a superposed set of emission lines of hydrogen and possibly of other elements.

However, as we have seen, the lifetime of a solar nebula is relatively short — perhaps less than 10^6 years. Even though many stars during their own long lifetimes may have been at some time surrounded by such a nebula, not many would be expected to be in this stage now.

To find a "solar nebula" belonging to another star, a search should probably be made among relatively young stars of about the mass of the sun. It is tempting to suggest that the so-called T Tauri variables are such stars. Their spectra show emission lines that are almost certainly produced in nebulous shells or envelopes. Their central stars have continuous spectra with absorption lines that often appear "veiled" as though the starlight is shining through a fairly thick obscuring layer. These absorption lines are usually broad, as though the central stars have large rotational velocities which could perhaps be explained in terms of a protostar that had not yet accomplished the separation of angular momentum required by McCrea's theory. And some T Tauri variables have apparently been forming quite recently in the Orion nebula, which has long been regarded as an enormous cauldron in which new stars are born even at the present time.

II. STELLAR EVOLUTION

The Milky Way is a vast spiral structure (Figure 21) consisting of a massive, almost spherical nucleus a few hundred light years in size, which is made up of a great number of stars, mostly cool dwarfs whose space density in its innermost regions is about 24,000 times greater than in the "local swimming hole." The average distance between two neighboring stars is only $\frac{1}{10}$ light year, instead of 4 light years. If we were located in this central region (Figure 22), our nearest neighbor, α Centauri, instead of having a visual apparent magnitude of zero would be seen as a star of apparent magnitude -8: it would be intermediate between Sirius and the full moon. The nucleus of the Milky Way rotates almost as a solid body around an axis at right angles to our line of sight, with a linear velocity of about 250 km/sec at its equator.

Outside the nucleus there is a flat structure of spiral arms resembling those of various other galaxies (Figures 23, 24). These consist of stars, gas, and dust, and there is also gas in the central nucleus. The "pinwheel" of spiral arms rotates around the galactic nucleus, roughly in accordance with Kepler's third law, which implies that the linear velocity should decrease with increasing distance from the center of the Milky Way, as $\sqrt{D}$, where D at the location of the sun is about 30,000 light years.

There are also stars between the spiral arms, as well as above and below the "galactic plane" defined by the spiral arms, but their space density is much smaller than in the "local swimming hole." The gas out of which the stars are condensed is also mostly concen-

Figure 21. Spiral galaxy M 51 (Whirlpool nebula), which resembles the Milky Way in structure; seen face-on.

(Mount Wilson and Palomar Observatories photograph with the 200-inch telescope.)

Figure 22. Photograph of the Milky Way made with the Henyey-Greenstein wide-angle camera. The center of the Milky Way is located beyond the obscuring clouds on the left side of the photograph. The southern Coal Sack is approximately in the center of the picture.

trated in the spiral arms; in the "local swimming hole" the average gas density is about 1 H atom per cm^3. In the galactic nucleus it is several hundred times greater. Between the spiral arms and in the "spherical halo" of the Milky Way, the gas density is probably of the order of 1 H atom per 100 cm^3.

Optical observations of the Milky Way show many regions of heavy obscuration by cosmic dust clouds (Figure 25). But the dust contributes only a small amount, of the order of 1 per cent, to the density of the gas. Nevertheless, it probably plays an important role as an

Figure 23. Spiral galaxy NGC 4565; seen edge on.

(Mount Wilson and Palomar Observatories photograph with the 200-inch telescope.)

intermediate step in the process of star formation. As far as we know, the dust clouds, like the hydrogen gas clouds, are strongly concentrated toward the galactic equator: the thickness of the disk is of the order of $\frac{1}{100}$ of its diameter.

Figure 24. Andromeda galaxy M 31. This galaxy is inclined about
70° to the tangential plane of the celestial sphere.

(*Lick Observatory photograph.*)

Figure 25. Milky Way in Sagittarius, Ophiuchus, and Scorpius. (*Yerkes Observatory photograph.*)

The galactic disk (Figure 26) is exceedingly flat in its inner regions: up to distances of the order of 20,000 light years from the center it does not deviate by more than about 200 light years from a plane. But at distances greater than about 30,000 light years from the center, the disk appears to be bent upward on one side and downward on the opposite side. According to F. Kahn and L. Woltjer, this may be due to the resistance which the entire galactic system experiences in its motion through an exceedingly tenuous intergalactic medium.

The Dutch astronomers have found that the cold hydrogen gas at a distance of about 10,000 light years from the center expands radially outward with a velocity of 50 to 100 km/sec, in the galactic

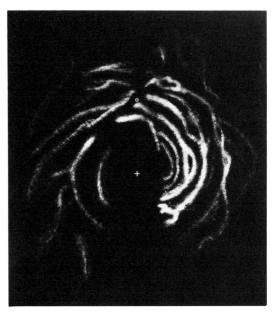

Figure 26. Spiral structure of the Milky Way, as deduced from radio observations of the 21-cm line of neutral hydrogen, made in Holland and Australia. The cross represents the galactic center, the circle the position of the sun. The inner portions of the Milky Way are not shown.

plane. They conjecture that at this rate the central regions of the Milky Way would be depleted in hydrogen in about 10^7 to 10^8 years. Since we know that there is a heavy concentration of hydrogen in these central regions, they suggest that it is continuously replenished by a stream of gas from the galactic halo: the gas diffuses into the halo at great distances from the center and streams into the central nucleus at its poles. There may thus be a continuous circulation of gas in the galaxy.

When we examine the apparent brightnesses of the stars in a cluster like the Pleiades (Figure 27), we immediately recognize that the intrinsic luminosities of the individual members are not the same. As a rule, the brightest stars in each cluster are either blue or very red (Figure 28). The faintest members of each cluster are always red. The spread in true luminosity between the brightest and the faintest objects may be as much as 100 million. This enormous disparity makes it immediately apparent that the luminosity of a star must be a particularly important quantity and one that is likely to give us considerable information concerning the physical properties of the stars.

Unfortunately, for the stars in general it is possible only in a very few cases to determine directly the true luminosities. In a cluster

Figure 27. The Pleiades star cluster and nebulosity.

(*Lick Observatory photograph.*)

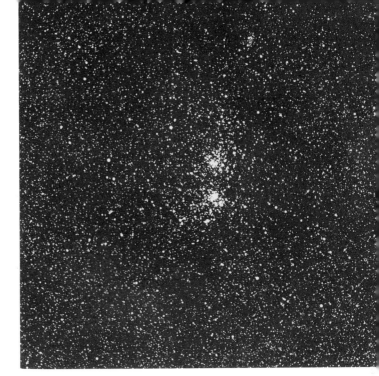

Figure 28. Double cluster in Perseus, a typical galactic cluster containing many blue stars.

(Yerkes Observatory photograph.)

we can arbitrarily assume as our unit of measurement the luminosity of the brightest star and refer to it the luminosities of all other stars, and the same can be done in the case of a double star. In order to determine stellar luminosities, not relative to some unknown standard but with respect to a known quantity, for example, the luminosity of the sun, we must make allowance for the differences in the distances of the stars. The latter are now known for a few thousand objects, so that for them, at least, we can compute the intrinsic luminosities in terms of that of the sun. It turns out that the most luminous stars are approximately one million times more luminous than the sun and the faintest stars are about one million times less luminous. The total range corresponds to a factor of 10^{12}.

The next significant parameter describing the properties of a star is its mass. This can also be determined only with difficulty. The methods now in use depend upon the universal law of gravitation as applied to double stars, either of the visual or of the spectroscopic type. Kepler's third law provides a relation between the period of a double star and the major axis of its orbit, and Newton's law of gravitation amplifies the relation in such a way as to bring in the mass of the object. In the case of a spectroscopic binary we can measure the orbital velocity and the period. From these two quantities we obtain the size of the orbit, in linear measure, and the masses of the two stars in terms of the mass of the sun. In practice, the problem is often more difficult. We must know the inclination of the plane of the orbit to the line of vision, and we must also measure the spectral lines of both stars of the binary in order to compute the individual masses of the two stars. But it is often possible to resolve these difficulties and to obtain trustworthy values. These range from about 50 times the mass of the sun to about $\frac{1}{10}$ or $\frac{1}{20}$. It is probable that even less massive stars exist as planet-like companions to such well-known objects as 61 Cygni and two or three other systems. Sarah Lippincott of the Sproul Observatory recently found that the fourth-nearest star, Lalande 21185, is accompanied by an unseen body whose mass is about $\frac{1}{100}$ that of the sun. Whether this is a self-luminous, though faint, star, or a dark planet like Jupiter is not known. The period of this binary is 8 years. But these small masses exert only a minute effect upon the motions of the heavier members of their systems, and the actual quantities involved are almost at the limit of measurement. Nevertheless, there is no reason to think that masses of the order of $\frac{1}{100}$ that of the sun may not be very frequent in our galaxy.

The third principal parameter in stellar astronomy is the radius of a star. In a few exceptional cases this quantity can be measured in angular units with the help of a large interferometer. But even the largest stars are not big enough to give us an accurate determination of their radii. For example, Antares has an apparent radius of only 0.02 second of arc. Since its distance is of the order of 300 light years, we find, according to F. G. Pease, that the true radius is approximately 150 times that of the sun. Much more reliable values of the radii can be obtained from the combination of photometric and spectroscopic observations of eclipsing variables. In principle, it is always possible from the duration of the partial and total phases of each eclipse and from the period of the variable to determine the radii of the two stars in terms of the distance between their centers. The spectroscopic observations permit us to express this latter quantity in kilometers; we can then also determine the sizes of the two stars in kilometers. Finally, there is a simple relation between the temperature and the radius of the star on the one hand and its luminosity on the other. Stefan's law of radiation gives the total amount of energy emitted by a square centimeter of the surface of an object as a known numerical constant times the fourth power of the temperature. Hence, if we know the temperature from the spectrum of a star, we find how much radiation is emitted by every square centimeter of its surface. But the total surface is $4\pi R^2$. If we multiply this by the radiation of each square centimeter, we obtain the total luminosity. Therefore if the total luminosity of the star, as known from its distance and brightness measurements, is divided by the radiation per square centimeter, the area of the total surface and the radius of the star are obtained.

As information concerning the radii, masses, and luminosities of the stars began to accumulate, astronomers noticed that not all combinations of the three parameters are realized in our galaxy. It is often convenient to represent each star as a point in a three-dimensional model in which the three coordinates are, respectively, luminosity, radius, and mass. Such a model was constructed several years ago at the Yerkes Observatory by G. P. Kuiper (Figure 29). In a square-cornered box the vertical dimension is taken to be the logarithm of the luminosity in terms of that of the sun. Thus when we read $\log L = +1$, the corresponding star has a true luminosity of ten times that of the sun. The mass is represented as the horizontal coordinate running toward the right and is also given in logarithmic measure, $\log M$. Zero means a star having the mass of the sun. Two would be a star having a mass one hundred times that of the sun. The radius is plotted as $\log R$ along the other horizontal scale running from left to right; -2 indicates a star whose radius is one hundred times less than that of the sun, and $+3$ is a star whose radius is one thousand times that of the sun. The stars are shown by means of beads suspended on thin wires in such a way that each bead has the correct values of $\log L$, $\log M$, and $\log R$. The beads are arranged

Figure 29. Three-dimensional model showing the distribution of stellar radii, masses, and luminosities.

(Yerkes Observatory photograph.)

in space as a fairly narrow band from the lower-left side of the model to its upper-right side. A white piece of string was used to show the region where the beads are most concentrated. This grouping is known as the main sequence. In addition to it there is a group of beads corresponding to very small values of L, M, and R. These stars are not a part of the main sequence but form a separate group known as the white dwarfs. Finally, there are a number of smaller groups: giants, supergiants, subgiants, and subdwarfs that are not very frequent in galactic space and are, therefore, represented by only a few objects in the model.

It is instructive to examine the actual model from different sides. If we look at it from the right, we see the beads projected against the narrow left-hand surface of the model whose coordinates are $\log L$ and $\log R$. The actual projections of the beads are shown by white dots pasted on a black surface. We see that in this projection the main sequence forms a fairly conspicuous straight line with a few stars departing from it on the right, and a group, distinctly removed, in the left-hand corner. The stars on the right side of the main sequence have larger radii and are different kinds of giants. The stars on the left are white dwarfs and subdwarfs.

If we examine the model from the left, as is done by the young lady in the picture, we observe the stars projected upon the right-hand side of the model whose coordinates are log L and log M. Here again the great majority project themselves as a narrow band running from the lower left to the upper right. This sequence is known as the mass-luminosity relation. It was first extensively treated by A. S. Eddington, although there had been earlier indications of it in the work of E. Hertzsprung and others. In this projection the white dwarfs again form a detached group. They do not agree with the mass–luminosity curve but fall below it; in other words, they have insufficient luminosities for their masses.

The left-side projection, which includes the main sequence, is almost identical with the famous diagram now known as the Hertzsprung–Russell, or H–R, diagram. It was first described by Henry Norris Russell in a lecture on December 30, 1913, in a joint session of the American Astronomical Society and the American Association for the Advancement of Science at Atlanta, Georgia. Following some earlier work by himself and by Hertzsprung, he had plotted the luminosities of the stars against their spectral types. The majority fell in a narrow band, the main sequence, with the rest of the stars forming a broader band running approximately horizontally from left to right, the giant sequence. The spectral types, O, B, A, F, G, K, M, and N, are essentially measures of the surface temperatures of the stars, running from about 20,000° at spectral class B to about 2,500° at class N. Because of the relation between L, T, and R, which we have already discussed and which is written $L = 4\pi R^2 \sigma T^4$, it is always possible to transform the original H–R diagram into another in which the horizontal scale is not the temperature but the radius. In this representation of the radius–luminosity relation, the main characteristics of the H–R diagram are retained, but the sequences are tilted in a slightly different manner. It is most astonishing that the sequences are so clearly marked, both in the L, R and in the L, M projections.

During the half century since Russell's lecture, a large amount of information has been accumulated concerning the parameters of the stars. We now know that the relations are statistical in nature; actually there are few, if any, parts of the space enclosed by our model which are "prohibited." For example, it was thought in 1913 that all blue and white stars had large luminosities and fell in the upper part of the H–R diagram. Soon afterward it became apparent that the companion of Sirius, though white in color, has a very small mass. It was the first clearly recognized member of the white-dwarf group. Later we began to notice that the so-called Wolf-Rayet stars, though very blue and hot, are not as luminous as are the normal O-type stars with which they are often combined in binary systems. Still more recently it was recognized, especially by Vorontsov-Velyaminov, that the ordinary novae, many years after their outbursts, are often blue stars of relatively small intrinsic luminosity. This led to the idea that another complete sequence of stars exists in the H–R diagram, running from top to bottom and joining the very luminous O-type stars with the white dwarfs.

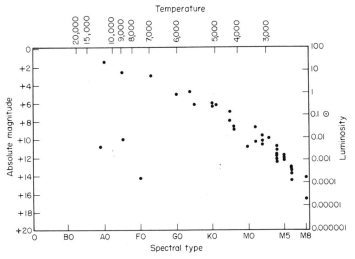

Figure 30. The H–R diagram for stars of the solar neighborhood whose distances are less than 15 light years.

It is not yet certain that this sequence has a real evolutionary significance, as some astronomers believe. But it must be recognized that representative points may be found in many parts of the octant shown in our model which were previously believed to be completely empty.

We should not confuse this result with the well-established statistical tendency of the stars of our galaxy to occur in preferential bands within the model (Figures 30, 31). It has been estimated that approximately 100 billion stars of the Milky Way system belong to the main sequence and obey the mass–luminosity relation. The total number of white dwarfs may be one hundred times less and the number of subdwarfs may be of the same order. The giants are

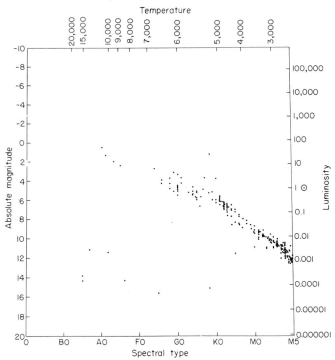

Figure 31. The H–R diagram for stars of the solar neighborhood whose distances are less than 30 light years.

probably ten thousand times less frequent than the main-sequence stars, and the supergiants are so rare that there may not be more than 1,000 or 10,000 of them throughout our entire galaxy.

An important quantity is the true spread of the main sequence at right angles to its extent. If we construct a model for a galactic cluster, we obtain an arrangement of the stars that runs almost exactly along a straight line through the octant, with the existing departures almost wholly accounted for by the remaining errors of measurement. This is illustrated by the H–R diagrams of the clusters Praesepe and the Pleiades (Figures 32, 33). The exceedingly small spread of the main sequence in the later spectral types of the Hyades has been confirmed by O. J. Eggen. It is accounted for by the fact that the stars in clusters have similar physical properties and were probably formed approximately at the same time. We cannot expect this kind of uniformity when stars are selected more or less at random.

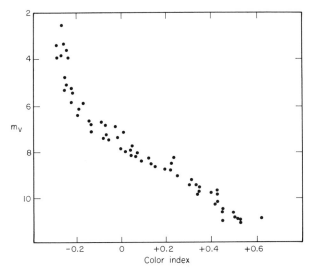

Figure 32. Color–luminosity array for the Pleiades, according to H. L. Johnson and W. W. Morgan. The temperature is represented by the color index of the star, along the abscissa. Zero corresponds to about 10,000°, or spectral type A0. The ordinate is the visual apparent magnitude, which is a measure of the absolute magnitude or luminosity.

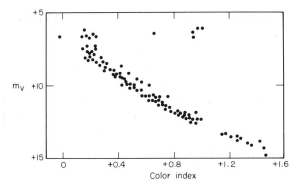

Figure 33. Color–luminosity array of the star cluster Praesepe according to H. L. Johnson. The coordinates are the same as in Figure 32.

For the galaxy in general the spread of the main sequence at a temperature of about 8,000°K is about two magnitudes. The principal cause of the spread is the chemical composition of the stars. A large amount of hydrogen, compared to other elements, results in a smaller luminosity than would be observed for a star consisting mostly of heavy elements. This relation between the chemical composition and the location of a star within the main sequence led some years ago to the hope that we would be able to determine accurately the abundance of hydrogen in stars and, consequently, their ages. However, more recent work has shown that the problem is not so simple. There are many objects that depart from the main sequence and the mass–luminosity relation for causes other than those connected with the chemical composition. For example, the companions of certain close double stars, like XZ Sagittarii, R Canis Majoris, and DN Orionis, are fairly luminous objects, despite the fact that their masses are only a fraction of the mass of the sun. These stars depart in a most conspicuous manner from the mass–luminosity relation. They are usually subgiants with fairly large radii, exceeding that of the sun by factors of from 2 to 5. But their masses, as we have seen, are often much smaller than the mass of the sun.

It has recently been emphasized, especially by A. J. Deutsch, that instead of using the conventional forms of the L, R and L, M diagrams we should make greater use of some diagrams which were introduced into astronomy years ago by R. Hess. In these, curves were plotted, representing lines of equal star frequency referred to a unit of volume in space near the sun. In this manner we should obtain not only the geometrical properties of the different sequences in each of the two projections of our model, but we could read off directly the probability of finding a star at a given point. Without a clear recognition of the stellar densities in space, we shall not have an adequate basis for studying the origin and evolution of the stars. For example, it must be cosmogonically important that along the main sequence the density increases from top to bottom, so that the red dwarfs are one hundred or one thousand times more frequent, per unit volume, than are the blue stars.

The model (Figure 29) applies only to the "local swimming hole" in our galaxy. In the central regions of the Milky Way and of extragalactic systems the distribution of the stars in our model would

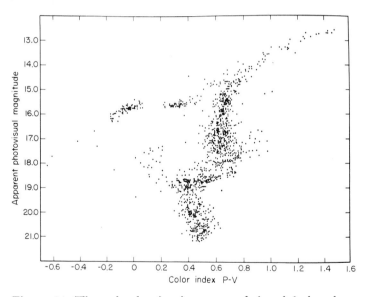

Figure 34. The color-luminosity array of the globular cluster M 3, according to H. C. Arp, W. Baum, and A. Sandage. The coordinates are the same as in Figure 32.

be different. This has led to the recognition of different types of populations in the universe. Long ago H. Shapley noticed that in globular clusters the giant sequence is quite different from that of our galaxy, and Baade has introduced the term Population II to describe the arrangement of the stars with respect to L, M, and R, in globular clusters and in the central nucleus of our galaxy (Figures 34, 35).

As a star evolves, it changes in radius, surface temperature, and luminosity. Whether it also changes in mass, other than through the conversion of mass into radiation, depends upon the degree of instability of the star. It seems reasonably certain that large changes in mass occur only in the latest stages of stellar evolution.

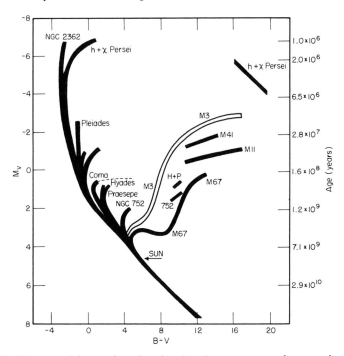

Figure 35. Superposition of color–luminosity arrays of several galactic clusters and of one globular cluster (M 3).

(*After A. Sandage.*)

At the beginning of its life the star is simply a condensation of interstellar gas and dust, large in size, relatively cold throughout ·its material, and low in density. The condensation is held together by its own gravitation and may be aided by the pressure of hot, ionized hydrogen from its surroundings. The condensation contracts, becomes hotter and denser inside. The representative point of such an object lies, at the start, far outside the limits of the H–R diagram on the right side. Gradually, however, the point describes an "evolutionary track" in the diagram, from right to left, and sloping slightly upward (Figure 36).

If we follow the evolutionary track of a particular star in the H–R diagram, we distinguish the following four distinct sections:

1. The stage of Kelvin contraction, from right to left, which may last (according to a summary by G. Herbig)

3×10^4 years for B0 stars of mass $20\odot$
2×10^6 years for A0 stars of mass $3\odot$
5×10^7 years for dG2 stars of mass $1\odot$
2×10^8 years for dK5 stars of mass $0.6\odot$
$\sim 10^9$ years for dM5 stars of mass $0.2\odot$

2. The "stable" main-sequence stage during which the star remains very close to what is often designated as the "zero-age" main sequence. This stage may last

$\sim 10^7$ years for B0 stars of mass $20\odot$
$\sim 5 \times 10^8$ years for A0 stars of mass $3\odot$
$\sim 10^{10}$ years for dG2 stars of mass $1\odot$
$\sim 10^{11}$ years for dK5 stars of mass $0.6\odot$
$\sim 10^{12}$ years for dM5 stars of mass $0.2\odot$

Although these time intervals are very uncertain, they are all about one hundred or one thousand times longer than the first stages.

3. The fairly rapid stage of exhaustion of hydrogen fuel in the stellar interiors during which the tracks move from left to right, slowly at first, and then more rapidly until the star becomes a relatively cool giant. I hesitate to give time intervals for the entire stage. But we may infer something about this from the initial theoretical tracks computed by various persons. Thus we can estimate from a diagram by Strömgren that a star of about $10\odot$ would move from B0 to B2 in roughly 2×10^7 years, and to B4 or B5 in

another 2×10^7 years. A star of $3.5\odot$ would move barely perceptibly (say, by one-half of a spectral class) in 10^8 years. The third stage, though probably somewhat shorter than the second, is considerably longer than the first. It seems to me, therefore, that we can easily account for the following facts:

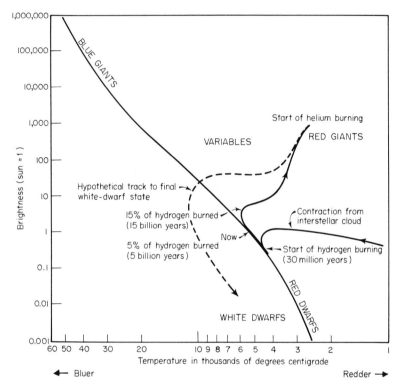

Figure 36. The evolutionary track of a solar-type star. The tracks of other stars resemble that of the sun: those of larger mass lie above, of smaller mass below the track of the sun. The contraction periods are relatively short. Each star remains for a relatively long interval of time on or near the main sequence, and then moves to the right side of the main sequence. In its last stage the track runs from right to left, and finally drops rapidly into the region of the white dwarfs.

(*a*) The great majority of stars whose masses are greater than $1\odot$ are close to the zero-age main sequence.

(*b*) Most of the fairly massive stars that lie slightly to the right of the main sequence must be in the third or fourth stages of evolution. Only very few massive stars can be expected in the first stage.

(*c*) Among the stars of small mass ($<1\odot$) many may still be in the first stage; this may be the case for the flare stars of the UV Ceti type (mass $\sim 0.4\odot$).

(*d*) Stars of the smallest known masses ($0.1\odot$ and less) may not have had time to reach the main sequence and may all be in the first stage (T Tauri variables, etc.).

4. In the fourth stage the track runs again from right to left, crosses the main sequence and then perhaps dips rapidly downward through the region occupied by the novae and nuclei of planetary nebulae, into the domain of the white dwarfs. Again, I hesitate even to guess the time intervals involved, but it seems to me from general principles that they must be fairly short, since they involve rapidly acting processes of contraction and nuclear reactions at very high temperatures. The number of stars observable in this stage is small, but apparently we can separate them by spectroscopic means from stars in the second and early third stages. I am referring here not only to novae and planetary nebulae but also to the hydrogen-poor, blue stars investigated by Münch, Greenstein, Aller, and others.

The sun is a normal dwarf star whose radius, $R_\odot = 7 \times 10^5$ kilometers, as determined directly from its angular radius, ¼ degree, and its distance, 1.5×10^8 kilometers. The solar mass, deduced from the annual motion of the earth and Kepler's third law is $M_\odot = 2 \times 10^{33}$ grams. The total energy output of the sun is obtained from the "solar constant," the amount of energy received by 1 cm² at the surface of the earth, 1.4×10^6 ergs/sec cm² (or 2 calories/minute cm²) and the number of square centimeters on a sphere whose radius is 1 a.u., $4\pi(1.5 \times 10^{13})^2 = 2.7 \times 10^{27}$ cm². The total energy generation of the sun is $1.4 \times 10^6 \times 2.7 \times 10^{27} = 4 \times 10^{33}$ ergs/sec.

Since the surface area of the sun is $4\pi(7 \times 10^{10})^2 = 6.2 \times 10^{22}$ cm², each square centimeter of solar surface (the photosphere) radiates

$$Q = \frac{4 \times 10^{33}}{6 \times 10^{22}} = 7 \times 10^{10} \text{ ergs/sec/cm}^2.$$

Stefan's law, $Q = 6 \times 10^{-5} T^4$, thus gives for the "effective" temperature of the solar surface

$$T_e = \sqrt[4]{\frac{7 \times 10^{10}}{6 \times 10^{-5}}} \sim 6,000°\text{K}.$$

Since the total energy output is 4×10^{33} ergs/sec and the mass of the sun is 2×10^{33} grams, 1 *average* gram of solar gas emits

$$E = \frac{4 \times 10^{33}}{2 \times 10^{33}} = 2 \text{ ergs/gm/sec}.$$

The sun must be at least as old as the earth, say, 3×10^9 years, or 10^{17} seconds. Hence each average gram has already radiated into space at least 2×10^{17} ergs, which exceeds by a factor of about 1 million any known chemical process of energy production. Even the energy generated by the gravitational contraction of the sun fails by a very large factor. We conclude that the solar energy comes from nuclear processes in its interior. The relevant nuclear processes, however, require very high temperatures — of the order of millions of degrees; we must, therefore, explain why we believe that such temperatures really occur in the interior of the sun.

A simple demonstration runs as follows: Imagine that a plane through the center divides the sun into two bodies, each containing 1×10^{33} grams, with their centers separated by approximately one solar radius, 7×10^{10} centimeters. The gravitational attraction between these two bodies is very nearly

$$F = G \frac{m_1 m_2}{r^2} = 6.6 \times 10^{-8} \frac{10^{33} \times 10^{33}}{(7 \times 10^{10})^2} = 10^{37} \text{ dynes}.$$

This force acts upon a surface of area

$$\pi (7 \times 10^{10})^2 = 2 \times 10^{22} \text{ cm}^2$$

and, therefore, exerts an average pressure of

$$P = \frac{10^{37}}{2 \times 10^{22}} = 5 \times 10^{14} \text{ dynes/cm}^2.$$

The solar gases obey the usual equation of state

$$P = \frac{k}{\mu H} \rho T,$$

where ρ is the average density of the sun, 1.4 gm/cm³, μ is the atomic mass and may be set equal to 1, and H is 1.7×10^{-24} gram. Therefore,

$$5 \times 10^{14} = \frac{1.4 \times 10^{-16}}{1 \times 1.7 \times 10^{-24}} \, 1.4T,$$

from which we find $T_\odot$ (average) $= 10^7$°K. More accurate calculations permit us to determine the distribution of T from 6,000°K at the photosphere to 20,000,000°K at the center. The corresponding density at the center turns out to be about 150 gm/cm³.

The mean velocity of a hydrogen atom at room temperature (300°K) is of the order of 3 km/sec. Since the velocity increases as $\sqrt{T}$, the mean velocity at 2×10^{7}° is of the order of 800 km/sec. The relative velocity of two protons colliding head on may be in many cases several thousand km/sec. It is easy to show that with such large velocities many colliding protons will overcome their electrostatic repulsion and cause the phenomenon of fusion. It is, therefore, reasonable to believe that the energy of the sun is derived mainly from the fusion of the most abundant element, hydrogen, to produce helium.

In this process four H nuclei fuse to produce one nucleus of He. A little less than 1 per cent of the mass of the four original protons is radiated away in the process. Since about one-half of the sun now consists of hydrogen, about 1 per cent of every ½ gram of solar gas will be ultimately converted into helium, giving us a total supply of energy:

$$E = \tfrac{1}{2} \times 0.01 \times c^2 = 0.5 \times 10^{19} \text{ ergs}.$$

Thus, if every gram were to partake in the process, the "life expectancy" of the sun would be

$$\frac{0.5 \times 10^{19} \text{ ergs/gm}}{2 \text{ ergs/(sec gm)}} = 2.5 \times 10^{18} \text{ sec},$$

or about 10^{11} years. This value must, however, be reduced, perhaps by as much as a factor of 2 or 3, because there is little mixing of the outer and inner layers of the sun. A life expectancy of the order of 3 to 5 times 10^{10} years is a reasonable guess. Even so, we conclude that the present age of the sun, about 5×10^9 years, is only about $\frac{1}{10}$ of its total lifetime.

The absolute bolometric magnitude of the sun (as it would appear at a distance of 10 parsecs or about 33 light years) is $+5$. The most massive stars have bolometric absolute magnitudes of the order of -10. They are thus 15 magnitudes more luminous than the sun, which means that their total energy generations are 10^6 times greater: 2×10^{39} ergs/sec instead of 2×10^{33} ergs/sec. But they also are more massive than the sun, perhaps by a factor of about 50. Accordingly, their life expectancies would be shorter by a factor of $10^6/50 = 20,000$. These stars would run through their entire life cycles in intervals of the order of only

$$\frac{4 \times 10^{10}}{2 \times 10^4} = 2 \times 10^6 \text{ years.}$$

In other words, a very massive star, like Plaskett's famous binary HD 47129, each component of which has a mass of the order of 50 or 75 solar masses, was "born" no earlier than 1 or 2 million years ago. It appears certain that if star formation has been going on in the galaxy during the interval from -5×10^9 years (roughly the age of the sun) to -10^6 years, it must be going on today.

The average velocity of a star in space with respect to our "local swimming hole" (or, better, our local standard of rest) is about 10 km/sec. In 10^6 years such a star would have traveled a distance of

$$10 \text{ km/sec} \times 3 \times 10^7 \times 10^6 \text{ seconds} = 3 \times 10^{14} \text{ kilometers}$$

or 30 light years. As the average distance of such a star, bright enough to be thoroughly explored, is about 1,000 light years, the proper motion (if the velocity happened to be at right angles to our line of sight) would have carried it across an angle of roughly $1°$: the star is still approximately where it was born. The close association of all very young stars with the spiral arms of the Milky Way and, especially with their bright and dark nebulae, is thus an indication of the physical conditions of those regions of space where stars are formed (Figure 37).

No such simple conclusion is possible with respect to the older stars. But it is of interest that B. Strömgren, the author of the foregoing argument, has recently succeeded in determining the average ages of several groups of stars, all much older than the very massive stars we have been discussing. Then by means of their space motions he has succeeded in deducing where in the Milky Way these older stars were formed. The results are not yet very consistent because of the great uncertainty in determining the galactic orbits of stars from their space motions, but in a general way one may state that star formation has been going on during the past few billion years, mostly in the spiral arms of the Milky Way.

The observational data upon which our conclusions regarding stellar evolution are based have thus far been of a statistical nature: we observe many different kinds of stars and nebulae, and we have reasons for believing that in the first three stages of each evolution-

Figure 37. A group of bright and relatively young stars in Ophiuchus and Scorpius, which are associated with dark and luminous clouds of dust and gas in which they were formed.

ary track, and probably also in the earlier parts of the fourth stage, the mass of a normal star changes very little. Conversion of matter into radiation causes a loss of less than 1 per cent; other causes, such as prominence activity or the ejection of corpuscular streams, are probably effective only in very massive stars. Hence, if we consider only stars with masses no greater than, say, ten times that of the sun, we shall probably be correct if we attribute to evolutionary changes the observed differences in surface temperature (or radius) and in the intrinsic luminosity (or energy production) of stars having similar masses. In other words, if we should find a star of solar mass, but of much lower surface temperature and of much larger radius than the sun, we would conclude that it must either be in the first, the contracting, or in the third, the expanding, stage of its evolution, or in the early part of the fourth stage. If we also know the luminosity of this star, we should be able to tell unambiguously on which particular branch it is located. But only the tracks of stage one and the beginning parts of track three are accurately known from detailed computations of the response of a huge gas sphere to the nuclear reactions occurring in its interior. The exact form of the latter stages of tracks three and four are known only approximately. Nevertheless, the observations provide many examples of stars located in different parts of their evolutionary tracks.

The youngest formations are probably the so-called "globules," small, dark condensations of diffuse interstellar gas and dust which are often seen projected against the luminous background of a large gaseous nebula. The smallest globules have circular outlines and are, therefore, probably spherical in shape. They are sufficiently massive and compact to resist the disrupting forces acting upon them, and they almost certainly are capable of contracting into real stars by their gravitation. Some of the smallest globules have radii of the order of 10^{12} kilometers, or about 1 million times that of the sun. They are thus vastly larger in size than is the solar system as measured by the distance of Pluto from the sun, but they are much smaller than the original cells of interstellar matter, which we discussed in the preceding lecture. The globules are dark because of the absorbing action of the dust within them. Their masses are not known but are probably similar to those of the stars.

Since the process of Kelvin contraction is rapid for stars of large mass, and slow for stars of small mass, we should expect to find many small dwarfs, which are already luminous — because the contraction generates heat — but are still in the first stage of evolution. Such stars have been found in large numbers by P. P. Parenago in the region of the Orion nebula and by M. F. Walker in several very young galactic star clusters (Figure 38). Many of these contracting stars are variable in brightness, and they are often referred to as T Tauri-type variables. All have masses smaller than that of the sun. One or two stars of fairly large mass, found in galactic star clusters, may also be in the contracting stage, but the evidence is not yet fully convincing.

Since stage two lasts a long time during which the star changes only imperceptibly in radius and luminosity, most stars are actually found on or near the main sequence of the H–R diagram. But since this stage is of shorter duration when the mass is large, we should expect that some of the most massive stars, which are also most luminous when on the main sequence, have had time to convert most of their internal hydrogen into helium and would be observed on the right side of the main sequence.

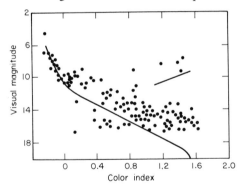

Figure 38. The color–luminosity array of the young galactic cluster NGC 2264, according to M. F. Walker. The heavy line represents the normal main sequences; the short line is the normal giant branch. The hotter stars are on the main sequence; the cooler ones lie above it.

The best evidence we have of this phenomenon comes from a study of the H–R diagrams of star clusters. These are more or less compact groups of stars, usually extending over a considerable range in apparent magnitude and spectral type (or temperature). The galactic, or open, clusters consist of several dozen or even several hundred stars, while the globular clusters contain hundreds of thousands of stars. All the members of one cluster may be thought of as having approximately the same age, but different clusters may have very different ages. The youngest clusters, like NGC 2362, which contain very massive and, therefore, very luminous stars, cannot be much older than their youngest members — a few million years in the case of NGC 2362. In this very young cluster, Walker found the stars of small mass to be still in their contracting stages of evolution. In the older galactic clusters, like the Pleiades and the Hyades, the H–R diagrams are lacking in hot, massive main-sequence stars — they must have gone through stages three and four of their tracks and are presumably now white dwarfs. These clusters are believed to have ages of the order of tens or even hundreds of millions of years. The oldest among the galactic clusters, like NGC 752, M 67, and NGC 188, have very few main-sequence stars with masses greater than that of the sun; their H–R diagrams are completely different from those of the younger clusters, except for the lower portions of their main sequences. The stars with masses between about 1 and 2 solar masses are found among the yellow and red giants. If they ever possessed stars of very large mass, they must have become white dwarfs a long time ago. The ages are of the order of up to 10^{10} years.

Finally, the very oldest formations in the Milky Way are the globular clusters whose gaseous constituents were swept away in the earliest stages of their evolution as a result of their passages across the plane of the Milky Way where their own gases collided with those of the spiral arms or the nucleus of the Milky Way. The members of the globular clusters form a pure type II Population — all their stars must be about 10 billion years old (Figure 39). The H–R diagrams of globular clusters are of a peculiar character: their main sequences break off at about the location of the sun; only short main-sequence stubs have been recorded for stars of

Figure 39. Globular star cluster in Hercules (M 13 = NGC 6205).

(Mount Wilson and Palomar Observatories photograph with the 200-inch telescope.)

small mass. All globular cluster are so far away that their faintest and, therefore, least massive members cannot be recorded with even our largest telescopes. Most of the more massive and luminous stars are located in a band that swings toward the upper-right side, forming a fairly sharp discontinuity in the region of the red giants. There is, however, another well-populated and nearly horizontal branch whose stars are about one hundred times more luminous than the sun. In a narrow stretch of this horizontal band we usually find a group of strictly periodical variables — the RR Lyrae stars — whose changes in brightness are caused by radial pulsations with periods of between about two hours and one day. It is probable that the stars of this branch are somewhat more massive than the sun: because of their more rapid evolutions they are already in the fourth stage and are moving from right to left with the passage of time. The very massive stars of the globular clusters must have become white dwarfs several billions of years ago. Because the latter are intrinsically very faint, and because the clusters are all very far away from us, we cannot see or photograph these white dwarfs in the globular clusters. But they must be fairly numerous.

If we are correct in assuming that an RR Lyrae star, as it grows older, moves from right to left along the horizontal branch, we should expect that its period would become shorter as it grows older. We can readily understand this because the luminosity of the RR Lyrae variable remains constant along the horizontal branch. But since the luminosity is the product of the star's surface area, and the amount of energy radiated by each square centimeter of this surface: $L = 4\pi R^2 \sigma T^4$; we infer that an increase of T (along the abscissa, from right to left) must be compensated by a decrease in the radius R. The star becomes smaller but retains a constant mass. Its density increases. But the period of a pulsation depends upon the mean density of the gas, just as the period of vibration of a string depends upon its tension. Expressed mathematically, we find that for any group of stars built on the same general model, $P^2\rho = $ constant, where $\rho = $ mass/volume is the mean density.

There appears to be some indication that, statistically at least, the RR Lyrae stars have shorter periods when they lie near the left edge of the variable star "gap" (the word gap is misleading: in preparing the H–R diagram of the clusters, the pulsating variables are usually omitted) than when they are on the right side. The horizontal branch is an evolutionary track, but this alone does not indicate whether the stars move from left to right (third stage) or from right to left (fourth stage).

If we could detect a systematic shortening in the periods of individual variables, we should be able to confirm our hypothesis fully. But this has not yet been possible: many variables have shown changes in period, but there are about as many periods that become shorter as there are those that become longer. In some the period may decrease for a few years and then, rather suddenly, start increasing.

The chances are that these changes in period are not of a simple evolutionary character.

There is, however, another group of pulsating variables in several of which slowly *increasing* periods have been detected. They are usually designated as the β Cephei, or the β Canis Majoris variables. Their brightnesses undergo periodic changes with amplitudes of a few hundreds or a few tenths of a stellar magnitude (a difference of one stellar magnitude represents a brightness ratio of 2.5, Figure 40). Their absorption lines change in wave length, which by the Doppler effect indicates that their outer layers periodically expand and contract. When these stars are plotted in the H–R diagram, they are found to occupy a band that departs appreciably from the normal main sequence of stars in stage two: Variables located near the bottom of the band have periods of the order of 3 hours, at the top the periods are of the order of 6 hours; from other considerations

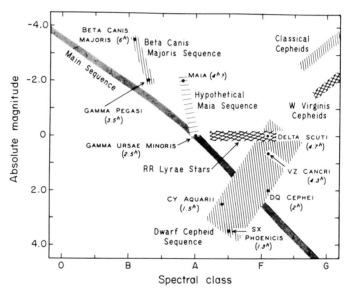

Figure 40. Location in the H–R diagram of several groups of pulsating variable stars.

we know that at the bottom of the band their masses are of the order of 5 $M_\odot$ and their radii are about 5 $R_\odot$. Variables at the top of the band have masses of 10 $M_\odot$ and radii of the order of 10 $R_\odot$. The mean density of the variables at the top are, therefore, $10/10^3 = 10^{-2}$ of the density of the sun, and those at the bottom are $5/5^3 = 5^{-2}$ of the density of the sun.

If we now write the fundamental relation between period and density for each group:

$$P^2\rho \ (\text{top}) \quad\ = c,$$

$$P^2\rho \ (\text{bottom}) = c,$$

and take the ratio, we expect to find that

$$\frac{P\ (\text{top})}{P\ (\text{bottom})} = \sqrt{\frac{\rho\ (\text{bottom})}{\rho\ (\text{top})}} = \sqrt{\frac{5^{-2}}{10^{-2}}} = 2.$$

The observations agree with this prediction, since

$$\frac{P\ (\text{top observed})}{P\ (\text{bottom observed})} = \frac{6\ \text{hours}}{3\ \text{hours}} = 2.$$

It is thus reasonable to assume that the fundamental relation $P^2\rho = c$ applies to the β Cephei stars as it is already known to apply to certain other groups of pulsating variables.

Several members of the β Cephei group belong to young star clusters (or associations). We must, therefore, assume that they are in the third evolutionary stage, not in the fourth, as was the case for the RR Lyrae variables. They have only recently left the normal main sequence (which is often, though misleadingly, designated as the "zero-age main sequence"), and their evolutionary tracks carry them from left to right (as is also the case for the brighter members of the Pleiades, Hyades, and other open clusters of intermediate age).

As they reach the left edge of the band, they begin to pulsate and continue doing so until they reach the right edge of the band. In the process of traversing the band, roughly horizontally from left to right, their surface temperature decreases but their luminosity remains the same. Consequently, their sizes increase and their densities decrease:

$$R = 6\,R_\odot \qquad R = 6.5\,R_\odot \quad \Big| \quad R\,7\,R_\odot$$

Band of β Cephei variables

Very roughly, the radius at the left edge is $6\,R_\odot$ and at the right $7\,R_\odot$. Since the mass remains the same, the ratio of the mean densities is

$$\frac{\rho\ (\text{right edge})}{\rho\ (\text{left edge})} = \left(\frac{6}{7}\right)^3 = \frac{1}{1.6},$$

and by the fundamental relation the ratio of the periods should be

$$\frac{P\ (\text{right edge})}{P\ (\text{left edge})} = \sqrt{1.6} = 1.26.$$

Near the middle line of the band the period of a typical β Cephei variable is about 17,000 seconds. At the left edge it would be about 15,000 seconds and at the right edge 19,000 seconds. There would thus be an increase in period of

$$19{,}000 - 15{,}000 = 4{,}000 \text{ seconds.}$$

We know only approximately how long it would take a star to traverse the band. But from the estimates given on page 54, we may assume an interval of about 10^7 years. Hence, if the period increases by 4,000 seconds in 10 million years, we might expect an increase of $(4 \times 10^3)/10^7 = 4 \times 10^{-4}$ second per year, or 0.04 second per century.

Can such a small increase in the period of a variable star be observed? If the period of the variable is 17,000 seconds, 1 year of 32×10^6 seconds would represent about $(32 \times 10^6)/(17 \times 10^3) = 2{,}000$ cycles. Within any single cycle we can determine the exact time of minimum (or maximum) light with a precision of perhaps about 1 minute or 60 seconds. After a lapse of 1 year the precision of the period would then be given by about $60/2{,}000 = 0.03$ second — about what we require. Therefore, if we observe the variable for a few years and find its mean period during that interval, and then for another few years, we should have no difficulty in detecting the theoretically predicted increase.

In reality, one of these variables, BW Vulpeculae, discovered by R. M. Petrie at Victoria, B. C., has shown an increase in period amounting to 3 seconds per century. Several other variables have shown smaller increases, but as yet none has been found to have a decreasing period. It should, however, be pointed out that the number of β Cephei stars known at present is quite small — about fifteen — and some of them have not been observed long enough to show a change in period. I should also mention that recent theoretical studies by P. A. Sweet and V. C. Reddish would make it appear less likely that the increasing period of β Cephei itself, roughly 1 second per century, is the result of evolution. The observed changes in period are about twenty-five times larger than those we have computed. This would imply a time interval of less than 1 million years, instead of 10 million years, for a variable to traverse the band. This the theoreticians do not like, and I must admit that their objections are well supported by what little we now know about the time scale of track three in the H–R diagram.

Nevertheless, I believe that prolonged studies of the periods of these and other groups of pulsating variables give us great promise that we shall ultimately discover the evolutionary process in a particular star.

Until we can detect this process, the astronomer faces the same problem that a hypothetical observer on Venus would if he had obtained a snapshot of the earth through a break in his almost permanent layer of clouds. This snapshot might conceivably show the presence on earth of human beings of different sizes, masses, and energies. He might then be tempted to construct a radius–energy diagram — the analogue of the astronomical H–R diagram, and also a mass–energy diagram — the analogue of the astronomical mass–luminosity relation. He might even find it attractive to attribute an evolutionary significance to these diagrams. (I am here using the word evolutionary not in the usual sense but rather to indicate the aging process of a human being.) A theory of life might even enable him to predict the rate at which a child grows and becomes an adult. He would probably realize that so complex an organism as man must have required a very long time to evolve. He would then be able to estimate roughly the number of persons who have had time to grow to maturity; but his snapshot would show him that only a much smaller number of adults are actually in existence. Hence, he might logically conclude that almost as many individuals disappear as are born in a given interval of time. Without ever having observed the process of dying, he would conclude that death does terminate the lives of men.

So also does the astronomer conclude that after a certain interval a star disappears from the upper parts of the H–R diagram and becomes a white dwarf. But neither the observer on Venus nor the astronomer on earth can feel completely certain about his conclusions as long as he has not detected the aging process in a particular individual.

We have seen that in astronomy 10 years should be sufficient to detect a change in period amounting to 0.3 second in a variable whose average period is about 17,000 seconds. This corresponds to about one part in 60,000, and this is also approximately the corresponding change in radius. Such a fantastically small change in the radius of a star is entirely beyond all direct or indirect methods of measuring the sizes of the stars.

For a very rough analogy, we shall adopt a time interval of 10^8 years for a star to run through the entire third stage of its evolution. We believe that it is possible to detect the evolutionary change in radius after 10 years of observing. This is one part in 10 million.

The life span of an average person is, say, 50 years. One part in 10 million of 50 years is $50/10^7 = 5 \times 10^{-6}$ year, or $5 \times 10^{-6} \times 3.2 \times 10^7 = 150$ seconds. The hypothetical observer on Venus would thus have to obtain a second snapshot of the earth 2½ minutes after the first if he would wish to match his wits with those of the astronomer. Is there any parameter measuring the aging process of man after an interval of a few minutes? If not, then the biologists may take comfort in the thought that the astronomers have not yet actually ascertained that the small increases in the periods of the β Cephei variables are really of evolutionary significance.

We are on somewhat firmer ground when we attempt to attribute an evolutionary significance to the outbursts of novae and supernovae in the last, the fourth, stages of their tracks. I shall discuss this at greater length in the fifth lecture.

III. GALAXIES

On April 26, 1920, there occurred at the National Academy of Sciences in Washington, a famous debate on the "Scale of the Universe." The speakers were Dr. Harlow Shapley, then an astronomer at the Mount Wilson Observatory (he later became the director of Harvard College Observatory), and Dr. Heber D. Curtis, an astronomer at the Lick Observatory (he later became the director of the Allegheny Observatory in Pittsburgh, Pennsylvania, and still later the director of the Observatory of the University of Michigan). Both astronomers had made extensive studies of the arrangement of the stars in the Milky Way but had reached strikingly different conclusions. After a lapse of nearly forty years, their papers, published in the *Bulletin of the National Research Council*, Vol. 2, Part 3, No. 11 (Washington, 1921), make one of the most fascinating accounts in the history of science. Even today astronomers argue about who was right, and their appraisals are strongly colored by their own interests in different aspects of the problem of galactic structure. Both scientists were able to back up their conclusions with formidable arrays of observational data that they themselves had secured. Both had found it impossible to accept the other's conclusions, and both were, to some extent, misled by the unreliability of the work of other astronomers or by their own incorrect interpretations of such results.

Two principal questions were to be resolved:

1. What is the size and the structure of our own Milky Way galaxy?

2. Are the great spirals "island universes," or, as we now call them, extraneous galaxies resembling the Milky Way, or are they intergalactic objects located fairly close to the solar system?

Shapley had made extensive studies of the globular star clusters and had found them to form a roughly spherical "skeleton" of the Milky Way, whose center, he believed, is about 50,000 light years away from the solar system, in the direction of the constellation Sagittarius. He also found that many globular clusters have distances of tens of thousands of light years from us, and he assigned to the diameter of our galaxy a value of about 300,000 light years. We now know that while Shapley's distances of the globular clusters were somewhat exaggerated — mainly because in 1920 he did not know of the existence of the effect of interstellar absorption by cosmic dust — and while the distance of the galactic center is now known to be only about 30,000 light years, he was essentially correct in attributing to the solar system an off-center position in the Milky Way. The diameter of the Milky Way is now often listed as 80,000 or 100,000 light years, one-third of the value given by Shapley in 1920, but this is a rather uncertain quantity because the Milky Way has no sharp edge, and Shapley's result may not be too far off if we include those spiral arms, galactic and globular star clusters, and gaseous clouds that are most remote from the galactic center.

However, Shapley erroneously concluded that "the evidence . . . is opposed to the view that the spirals are galaxies of stars comparable with our own. In fact, there appears as yet no reason for modifying the tentative hypothesis that the spirals are not composed of typical stars at all, but are truly nebulous objects." In reaching this conclusion, Shapley was most impressed with three, then very recent, results: (1) Seares's deduction that none of the known spirals has a surface brightness as small as that of our galaxy; (2) Reynolds' study of the distribution of light and color in typical galaxies, from which he concluded that they cannot be stellar systems; (3) van Maanen's measurements of the proper motions of condensations in the spiral M 33, which had indicated a relatively fast rotation of that object around its center.

We now know why the surface brightness of the Milky Way, as seen from the earth, is fainter than the surface brightness of the more conspicuous spiral arms of the Andromeda galaxy and of M 33: The solar system is located in one of the more distant spiral arms of the Milky Way, in which the surface brightness is comparable to the surface brightness of the most distant, barely visible, arms of the Andromeda galaxy. Because of the heavy obscuration by dust clouds, we never see the inner spiral arms of the Milky Way, whose surface brightness is now known from radio astronomical observations to resemble the surface brightness of the inner arms of other galaxies.

Reynolds' results have not been confirmed by more recent studies. We now know that the light and the color of a typical spiral (Figure 41) can be fully explained in terms of the combined radiation of many billions of stars (Figures 42, 43).

Finally, the results of van Maanen were later found to have been produced by instrumental errors. The spirals are all so far away from the sun that no proper motions of stars (that is, motions expressed in seconds of arc per year, at right angles to the line of sight) or star clusters, located in them, can be measured.

Curtis was certainly correct in his defense of the "island universe" hypothesis. He attributed to the Andromeda galaxy a distance of 500,000 light years (the correct value is approximately 1½ million

Figure 41. NGC 5364, spiral nebula in Virgo.

(*Mount Wilson and Palomar Observatories photograph with the 200-inch telescope.*)

Figure 42. NGC 3031, spiral nebula in Ursa Major. Messier 81.

(Mount Wilson and Palomar Observatories photograph with the 200-inch telescope.)

Figure 43. NGC 147, galaxy in Cassiopeia, shows resolution into stars. Member of local group. Taken in red light.

(Mount Wilson and Palomar Observatories photograph with the 200-inch telescope.)

light years), and he estimated that the more distant spirals could be as far away as 100 million or more light years. He did not accept Shapley's earlier estimate of 20,000 light years for the Andromeda galaxy, because in that case "the minute spirals would need to be at distances of the order of 10 million light years, or far outside the greater dimensions postulated for the galaxy by Shapley."

Curtis pointed out that the spectra of spirals could only be explained by the integrated light of a large number of solar-type stars. He argued that the absence of spirals near the galactic equator could be easily explained by assuming the presence of an obscuring ring of diffuse matter surrounding the Milky Way and resembling the dark rings often observed in spirals that are seen edgewise. He insisted that in terms of Shapley's hypothesis "their abhorrence of the regions of the greatest star density (the galactic equator of the Milky Way) can only be explained on the hypothesis that they are, in some unknown manner, repelled by the stars." He considered it improbable that there is in existence such a repelling force.

He pointed out that the spirals have enormous radial velocities, as had been measured by V. M. Slipher and others from the Doppler displacements of the absorption lines. Curtis suggested that such large motions were never found in the Milky Way and that, therefore, the spirals are probably extragalactic objects resembling our own galaxy in structure and in size. He admitted that van Maanen's proper motions, if they were real, would indicate distances of the order of only a few tens of thousands of light years, but he concluded: ". . . in view of the hazy character of the condensations measured, I consider the trustworthy determination of the . . . proper motions (in spirals) impossible by present methods without a much longer interval than is at present available."

The recent work on galaxies has confirmed Curtis' conclusions except that he had underestimated the distances of the spirals by a factor of about 3.

Curtis was, however, wrong when he attempted to show that Shapley's distances of the globular star clusters and the resulting diameter of 300,000 light years for the Milky Way were in error by a factor of 10. He stated, "I hold therefore to the belief that the galaxy is probably not more than 30,000 light years in diameter." And he adhered to the view of Kapteyn and others that the solar system is located near the center of the Milky Way. He reached this conclusion because he incorrectly assumed that most of the bright stars in globular clusters are yellow and red dwarfs, like the sun, and he also discarded Shapley's period–luminosity relation for the pulsating variable stars of the δ Cephei type. On the whole, it is probably fair to say that each scientist was wrong about half the time and correct the other half. Our present knowledge of the galaxies is essentially based upon Shapley's conclusions in regard to the globular clusters and Curtis' views concerning "island universes" (Figures 44, 45).

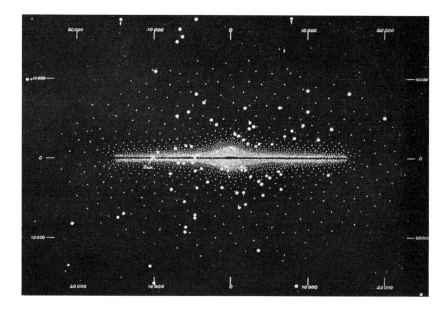

Figure 44. Structure of the Milky Way according to modern investigations (Shapley and others). The units of distance are parsecs: 10,000 parsecs are approximately 30,000 light years. The large spots are globular star clusters; the small dots are stars.

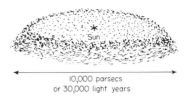

Figure 45. Structure of the Milky Way according to Kapteyn and Curtis.

The unaided human eye can distinguish between 2,000 and 3,000 stars scattered over the visible hemisphere on a moonless night. With an average binocular, the number can be increased to 10,000. The number of stars seen on photographs taken with the largest existing optical telescopes is between 2 and 3 billion. Hence, the star images on direct photographs are seen to be crowded together to such an extent that about 10,000 may be found within an angular area equal to that of the full moon. However, the impression of great star density is misleading. In reality, the individual stars of the Milky Way are separated by distances of several light years. If a model of the Milky Way were constructed with a scale such that the size of an average star is equal to that of a raindrop, their distances apart would have to be approximately 40 miles to give the right idea of the density of stars in a typical galaxy. For every cubic inch of stellar matter there are about 10^{22} cubic inches of almost empty space.

The average distances between neighboring galaxies are of the order of 1 million light years, and the space between the galaxies is even more nearly empty than is the interstellar space within a single galaxy. In the observable part of the universe there are about 10^{28} cubic inches of almost empty space for every cubic inch of stellar material.

The sun is an average star in brightness, mass, and size. At its distance from us of 1 astronomical unit, or 150 million kilometers, we describe its "apparent magnitude" by the number -27. A step of 1 magnitude in brightness represents a ratio of 2.5 in light intensity. If the sun were 30 light years away from us, it would appear to our eyes as a faint, naked-eye object of apparent magnitude $+5$. We use the term "absolute magnitude" to designate the brightness of a star if its distance were 30 light years. Our nearest stellar neighbor is α Centauri, whose distance is 4 light years and whose apparent magnitude is about zero. If α Centauri were 30 light years away, the intensity of its radiation would be 7.5×7.5, or about 56 times fainter than it is in reality. This results from the fact that at 30 light years it would be 7.5 times farther away than it is in reality and because the intensity of light diminishes as the square of the distance. But the ratio of 56 in the light intensity corresponds to a difference of about $4\frac{1}{2}$ magnitudes on the astronomical brightness scale. Hence, the absolute magnitude of α Centauri is approximately $+4.5$, which is similar to that of the sun.

Since we cannot photograph the most distant faint stars of the Milky Way, we must find an indirect method of determining the total number of stars in our galaxy. The answer comes from our knowledge of the velocity of rotation of the stars in the solar neighborhood around the distant galactic center in the constellation Sagittarius. When we measure the Doppler displacements of the spectral lines of distant galaxies located toward the constellation Cygnus and those near it, we find that their motions tend to be approximately 300 km/sec of approach. This is approximately the radial velocity of the Andromeda galaxy. In the opposite part of the sky the radial velocities of the galaxies tend to average 300 km/sec of recession. An example is the large Magellanic Cloud, whose radial velocity is approximately $+376$ km/sec. The direction toward Cygnus is approximately 90° from the direction toward the constellation Sagittarius, and therefore, we infer that we ourselves and all the stars in our vicinity revolve in an approximately circular orbit around the center of the Milky Way, with a velocity of 300 km/sec toward Cygnus. This would imply that the motions of the nearby galaxies tend to average about zero although, of course, every individual member of the local group of galaxies has a random motion of its own of several hundred km/sec. A similar circular velocity of rotation for the solar neighborhood stars is obtained from the radial velocities of the globular star clusters that form a spherical system without any appreciable amount of rotational flattening. This system of clusters does not participate in the phenomenon of galactic rotation; hence, the globular clusters tend to give large negative radial velocities in the vicinity of Cygnus and large positive velocities in the opposite part of the sky.

Since we already know that the distance of the sun from the galactic center is about 30,000 light years, or 3×10^{22} cm, we find that the period of one galactic revolution at a distance of 30,000 light years from the center is given by the ratio of the circumference of the circular orbit $2\pi \times 3 \times 10^{22}$ divided by the velocity 3×10^7 cm/sec. The result is 200 million years.

However, the Milky Way does not rotate as a solid body. Except near the center its motions obey the third law of Kepler, which states that the product of the mass of the galaxy, in terms of the mass of the sun, multiplied by the square of the period, expressed in years, is equal to the cube of the radius of the orbit, expressed in astronomical units. Since the period and the radius of the circular orbit are known, we find by simple arithmetic that the mass of the galaxy, or better, the mass of that part of the galaxy that is closer to the galactic center than 30,000 light years, is approximately 200 billion times the mass of the sun. Expressed in other words, this means that the number of stars in the galaxy is approximately 200 billion.

When Kepler's third law is applied in a reverse manner by inserting in it the mass of the galaxy, 200 billion suns, and computing the periods that would correspond to different distances from the galactic center, it is found that these periods become longer and longer as we go to greater and greater distances beyond that of 30,000 light years. In the inner parts of the Milky Way, at distances of the order of 20,000 and 15,000 light years from the center, the orbital periods are shorter than 200 million years. The innermost parts of the Milky Way cannot be observed with optical instruments because, as we have already seen, they are hidden from our view by obscuring dust clouds. They can, however, be detected with radio telescopes, and recent observations, made mostly with the large radio instruments in Holland and in Australia, have shown that the innermost parts of the Milky Way do not obey in their motions the third law of Kepler but revolve around the center as a solid disk with a constant period.

Figure 46. NGC 4594, spiral galaxy in Virgo, seen edge on. Messier 104.

(Mount Wilson and Palomar Observatories photograph with the 200-inch telescope.)

The rotational motions of other galaxies have been determined mostly by N. U. Mayall, at the Lick Observatory, from measurements of the Doppler displacements of the spectral lines. Except in those galaxies that are seen edgewise (Figures 46, 47), our line of sight is not seriously cut off by obscuring clouds. In the Andromeda galaxy, for example, which is tilted approximately 70°, not only the spiral arms near the center but even the nucleus of the galaxy itself can be photographed. It turns out that in all cases the innermost part of the galaxies tends to revolve around the center as a solid disk and that only at distances of some 20,000 light years from the center do the velocities fall off with increasing distances, in accordance with Kepler's third law. Since the distances of the nearer galaxies can be reliably determined by means of an improved version of Shapley's period–luminosity relation of pulsating variables, the angular distances of the different spiral arms, star clusters, and nebulae in these galaxies can be expressed in terms of light years and so, of course, can also the over-all diameters of these relatively close galaxies be found. The results of these investigations indicate that the Andromeda galaxy and several others resemble the Milky Way in size, mass, and structure. The vast majority of the galaxies are, however, considerably smaller and less massive than these giant spirals.

We have already seen that the absolute magnitude of the sun — its brightness at a distance of 30 light years — is $+5$. It is of interest to compute the integrated absolute magnitude of all 200 billion stars contained in the Milky Way. Remembering that a difference of 1 stellar magnitude corresponds to a brightness ratio of 2.5, we infer that 200 billion solar-type stars taken together would be about 26 magnitudes brighter than is the absolute magnitude of the sun. Hence, the integrated absolute magnitude of the Milky Way is $+5 - 26 = -21$. A more exact calculation gives an absolute magnitude of about -20.

Figure 47. NGC 4244, spiral galaxy in Canes Venatici, seen edge on.

(Mount Wilson and Palomar Observatories photograph with the 200-inch telescope.)

What is the greatest distance at which we should still be able to recognize a giant galaxy similar to our own? The faintest apparent magnitude, that is, at the very limit of the most sensitive photoelectric recorders in the focal plane of the 200-inch Palomar telescope, is about $+24$. Since the absolute magnitude is -20, we find that at the edge of the observable part of the universe such a galaxy would be 44 magnitudes fainter than is its absolute magnitude, the latter referring to a distance of 30 light years. But 44 magnitudes correspond to a brightness ratio of 4×10^{17}. Applying again the inverse square of the distance law for the propagation of light, we then conclude that at the edge of the observable part of the universe the distance of the galaxy would be $\sqrt{4 \times 10^{17}}$, or about 6×10^8 times greater than is the standard distance of 30 light years at which the absolute magnitude is measured. The product of 30 light years times the factor 6×10^8 would give us a radius of the observable part of the universe equal to about 20 billion light years.

But in making this estimate we have overlooked the phenomenon of the recession of the galaxies (Figure 48). Among the local group of galaxies whose distances are of the order of a few million light years, this phenomenon has not been found, but at greater and greater distances all galaxies tend to move away from each other, with a radial velocity that appears to be proportional to their distances from each other. For example, a cluster of galaxies in the constellation Ursa Major, whose distance is approximately 400 million light years from us, is receding with a velocity of 42,000 km/sec. Even greater velocities of recession of the order of one-third, or perhaps one-half, of the velocity of light have been measured among the most distant galaxies known at the present time. The motion of a luminous object away from us implies a red shift of all the waves emitted by the object and an attenuation of the radiation that is received on the earth: the light of a very distant galaxy is redder and fainter than it would be if there were no expansion in the universe. We cannot now accurately correct our estimate of the radius of the observable part of the universe for the effect of the red shift, but a rough guess is approximately 10 billion light years. At still greater distances the radial velocities of recession of the galaxies

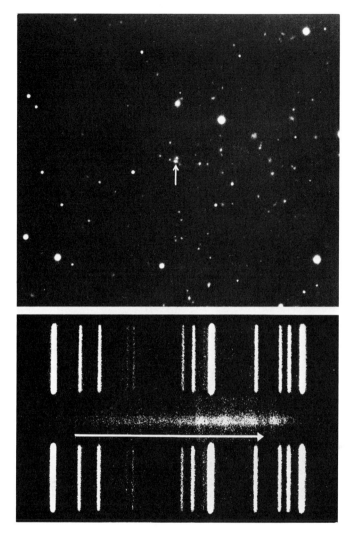

Figure 48. Hydra cluster of galaxies, showing galaxy observed for velocity, and spectrum showing large red shift (indicated by length of arrow) equivalent to a velocity of 38,000 miles per second.

(Mount Wilson and Palomar Observatories photograph with the 200-inch telescope.)

would approach the velocity of light, and of course, no radiation would ever reach us from a galaxy whose velocity is 300,000 km/sec.

Crude as our estimate may be of the radius of the observable part of the universe, let us assume that it is 10 billion light years and estimate the number of galaxies contained within that volume. Although the galaxies tend to appear more often in groups like clusters and even in superclusters of clusters, the average number of galaxies is about three per cubic megaparsec, where the word mega stands for 1 million and parsec for 3.3 light years. A simple arithmetical calculation then leads to approximately 100 billion galaxies within the observable part of the universe. Since our own galaxy is exceptionally large, it might be reasonable to assume that an average galaxy contains approximately 10 billion stars. The total number of stars in the observable part of the universe is then the product of the number of galaxies, 10^{11}, and the average number of stars in each, 10^{10}, or 10^{21}.

One of the most interesting recent results in the study of galaxies is the determination of the probability of their collisions. We have already seen that collisions of individual stars (page 24) within a galaxy are exceedingly rare events. Because of the much greater distances between two neighboring galaxies we might expect that collisions of galaxies are even less frequent. But this is not true. The average distance between two galaxies is of the order of 2×10^6 light years, within the general field, outside of fairly dense clusters or groups of galaxies. We may assume that their relative velocities are of the order of 1,000 km/sec. Then using the same method as was used on page 24, we find that no collision can take place during an average interval of

$$\frac{2 \times 10^6 \times 10^{18} \text{ centimeters}}{10^3 \times 10^5 \text{ cm/sec}} = 2 \times 10^{16} \text{ seconds} = 10^9 \text{ years.}$$

Since the solid angle subtended by a neighboring galaxy is about 4 square degrees, and since there are about 40,000 square degrees on the entire sphere, we conclude that the probability of one collision after 10^9 years is $1/10^4$, or we may say that one particular galaxy has a fair chance of colliding with another once in every 10^{13} years.

Thus, for any particular galaxy, like our own, the probability is exceedingly low that it has experienced even a single collision during the entire lifetime of the galaxy, say 6×10^9, or even 10^{10} years. But the number of observable galaxies is so great that a considerable number of them must have had such collisions. In a dense cluster of galaxies the probability of collision is so great, according to Spitzer, Baade, Minkowski, and others, that many, perhaps most, of their member galaxies may have experienced collisions with other member galaxies.

The duration of a collision can also be estimated. If we assume that the disk of one galaxy having a diameter of 100,000 light years passes at right angles through the disk of another galaxy, the duration of the entire collision would be

$$T = \frac{10^5 \times 10^{13} \text{ km}}{10^3 \text{ km/sec}} = 10^{15} \text{ seconds} = 3 \times 10^7 \text{ years},$$

provided that the relative velocity is 10^3 km/sec. It is thus reasonable to believe that our photographs should show a number of pairs of galaxies that are now in the process of collision; Baade and Minkowski have found several such pairs, and Zwicky has described many other pairs which have greatly distorted spiral arms and other peculiar structural features that suggest the action of strong — perhaps tidal, perhaps magnetic — forces (Figure 49).

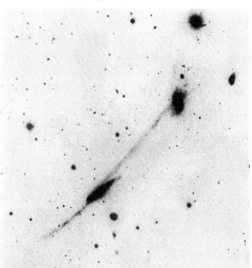

Figure 49. A pair of interacting galaxies.

Apparently the stars of the two galaxies do not collide head on, but their galactic orbits and, consequently, their spiral arms are greatly perturbed. The gases and dust clouds of the two galaxies do, however, collide head on; the diffuse media are arrested in their motion and are swept out of the galaxies. Intergalactic space must contain the remnants of these clouds, and there are indications that the existence of diffuse intergalactic matter makes itself felt in certain radio astronomical and optical observations.

The most fascinating problem of the universe is the explanation of the phenomenon of the red shifts of the distant galaxies. We cannot claim that this problem has been solved. The principal difficulty consists in the determination of the distances of the more remote galaxies. For the nearby galaxies Shapley's method, based upon the period–luminosity relation of the Cepheid variables, can be used: their stars are sufficiently luminous to be recorded on a direct photograph. Since the period of the variable is strongly correlated with its absolute magnitude, we need only find the period of a particular variable in the Andromeda galaxy and read off the curve the corresponding absolute magnitude. For example, a variable in our own galaxy whose period is 10 days was found to have an absolute magnitude (at 30 light years) of -4. The apparent magnitude of a 10-day variable in the Andromeda galaxy is $+20$. The difference between the apparent and absolute magnitudes $-24-$ corresponds to a ratio in light intensity of 4×10^9. Since the brightness of a star is inversely proportional to the square of its distance, we infer that the distance of the variable in the Andromeda galaxy and, hence, the distance of the entire galaxy, exceeds the standard distance of 30 light years by a factor of $\sqrt{4 \times 10^9} = 6 \times 10^4$. The distance of the Andromeda galaxy is, therefore, approximately

$$30 \times 6 \times 10^4 = 1.8 \times 10^6 \text{ light years.}$$

This method gives excellent results for nearby galaxies, but it fails at very great distances because the variables are then too faint to be seen as individual stars on even the best photographs. The most distant galaxies are, in fact, seen only as tiny nebulous spots, without any distinct structure, and at the limit of the sensitivity of the photographic emulsions. If it were possible to assume that all galaxies are roughly of the same linear size, we could obtain their relative distances from measurements of their angular diameters. But the galaxies differ enormously in size. Ours and the Andromeda galaxy are about 100,000 light years in diameter, but others, like NGC 185, have diameters of only 3,000 light years: a galaxy may appear small and faint because it is intrinsically a small object and not necessarily because it is very far away.

The best that we can do at the present time is to concentrate upon the rich clusters of galaxies, some of which contain hundreds or even thousands of member galaxies. We can then assume that the most luminous members of such a cluster resemble the largest galaxies of our neighborhood – the Milky Way itself and the Andromeda galaxy – and that the absolute magnitude is the same, about −20, and compute the distance with the help of the observed apparent integrated magnitudes.

There is some hope that a radio astronomical method of finding the distances of remote individual galaxies will soon be available. D. S. Heeschen, at the National Radio Astronomy Observatory, has measured the brightnesses of a group of relatively nearby galaxies in a certain frequency and has established with the help of their known distances the "absolute radio-frequency magnitudes." He has also measured the differences in brightness at different frequencies and has thus established their "radio-frequency colors." A plot of the absolute magnitudes against the color is the radio-frequency counterpart of the H–R diagram. The individual galaxies form in this diagram two distinct sequences. It should then be possible to measure the radio-frequency colors of more distant galaxies and read off Heeschen's diagram their radio-frequency absolute magnitudes – provided we can decide beforehand to which of the two sequences the galaxy belongs. If the "radio-frequency apparent magnitudes" have also been measured, the distances may be computed in the usual way.

We have seen that the galaxies differ greatly in size and luminosity and, therefore, also in mass. But do they all consist of the same kinds of stars and in the same relative proportions as in the Milky Way? And are the galaxies of about the same age? Answers to these questions should have an important bearing upon the theories of the origin of the galaxies and the cosmology of the universe. Important recent papers dealing with these questions have been published by W. W. Morgan of the Yerkes Observatory and N. U. Mayall of the Lick Observatory.

It is only since the construction of the large reflecting telescopes at Mount Wilson Observatory that our ideas about galaxies have been clarified. Visual observations could not give their true nature, but during the first part of the twentieth century long-exposure photographs resolved into stars the outer parts of the nearest spiral nebula — M 31 in Andromeda — showing that object to be a counterpart of our Milky Way system.

Because some of these stars in M 31 turned out to be Cepheid variables, the distance to the Andromeda galaxy could be determined. In 1929 Hubble found that M 31 Cepheids are about 4.6 magnitudes fainter than those Cepheids of the same period in the Small Magellanic Cloud. This placed M 31 at 8.3 times the distance of the Small Cloud. The latter is about 6° across, while the main part of M 31 is about $3\frac{1}{2}°$. Thus, if we could bring M 31 as near to us as the Small Cloud, it would be $8.3 \times 3\frac{1}{2}$, or about 30° in diameter — large enough to cover the Big Dipper. Evidently the Small Cloud is actually much smaller than the Andromeda system.

A few years ago in South Africa, A. D. Thackeray discovered numbers of cluster-type variables in the Small Magellanic Cloud. These rapidly pulsating stars have apparent magnitudes around 19.0, corresponding, after allowing for the dimming by interstellar dust, to a distance of about 160,000 light years. This is five or six times the distance of the sun from the galactic center; consequently, the Magellanic Clouds are independent galaxies, not merely condensations of the Milky Way. And M 31, now considered about ten times as distant as the Magellanic Clouds, must indeed be another system of stars like our own.

These two spirals are very similar, not only in size but in content. Both contain dark areas of absorbing material between the stars, emission nebulae, globular and open star clusters, and variable stars including novae and supernovae. Moreover, M 31 is also rotating around an axis at right angles to its plane of symmetry. But our system and M 31 are types of galaxies that differ greatly in size and shape from the Magellanic Clouds.

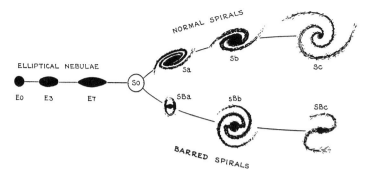

Figure 50. The sequence of nebular types according to E. Hubble. The diagram is a schematic representation of the sequences of classification. A few nebulae of mixed types are found between the two sequences of spirals. The transition stage, S0, is more or less hypothetical. The transition between E7 and SB_a is smooth and continuous. Between E7 and S_a, no nebulae are definitely recognized.

(E. Hubble, *The Realm of the Nebulae*, Yale University Press, New Haven, 1936, page 45.)

The classification of galaxies was attempted many years ago by Hubble, who recognized three basic varieties: S, spiral; I, irregular; and E, elliptical (Figures 50, 51). Among the spirals two main groups are distinguished – normal and barred (Figure 52). Each of these is subdivided into three classes, a, b, and c, depending partly upon the amount of material in the arms relative to that in the nuclear region, and partly upon the degree of openness of the arms and their resolution into stars. Both M 31 and our Milky Way galaxy are of the type Sb, whereas the Magellanic Clouds are called irregular (though thought by some to be barred spirals).

The arms of all spirals contain highly luminous O and B stars, as well as dust and gas. Baade believed that if dust and gas were lacking during the formation of a galaxy, spiral arms would not develop. Therefore, spiral arms seem to require the presence of dust and gas. But if the latter material were removed after the arms had formed (for instance, by collision with another galaxy), the spiral form would be preserved, but the galaxy would then consist only of stars.

Figure 51. NGC 3190, group of four galaxies in Leo. NGC 3185, type SB$_a$; NGC 3187, type SB$_c$; NGC 3190, type Sb; NGC 3193, type E2.

(*Mount Wilson and Palomar Observatories photograph with the 200-inch telescope.*)

Figure 52. NGC 7741, barred spiral galaxy in Pegasus.

(*Mount Wilson and Palomar Observatories photograph with the 200-inch telescope.*)

Irregular galaxies, typified by the Magellanic Clouds, lack a central nucleus and a plane of rotational symmetry. Hubble estimated that one galaxy in forty is an irregular one. In the Large Cloud blue stars and red supergiants are present, as well as obscuring dust which blots out background galaxies. Both clouds contain much neutral hydrogen gas, as Australian radio observations have shown. There are also several thousand typical Cepheid variable stars. All of these features are found in the spiral arms of the Milky Way and of M 31, and are representative of what Baade many years ago named Population I.

But the Large Cloud contains several globular clusters, in at least one of which are cluster-type variables of Baade's Population II. Hence, although the Large Cloud is mostly of Population I, it has a small admixture of Population II. The Small Cloud, however, is almost free of obscuring dust, for distant galaxies can be seen through it, undimmed. This suggests that it is richer in Population II objects than is the Large Cloud, but it does have a few hydrogen emission nebulae, characteristic of Population I.

Elliptical galaxies have a distinctive property. The shape of any one of them remains unchanged in a series of photographs taken with longer and longer exposure times. More stars in the outermost regions are recorded, but the curves of equal star density retain the elliptical contour shown on short exposures. Ellipticals are basically of Population II.

One well-known system of this kind is NGC 205, a companion of the Andromeda galaxy. Long-exposure photographs resolve it into stars, the brightest of which are red giants. Baade has found in this system two absorption patches, indicating dust, and a dozen luminous B-type stars. Except for these impurities, the predominant stellar objects in NGC 205 are of Population II.

Photoelectric measurement of the colors of various classes of galaxies gives a clue to their contents. J. Stebbins and A. E. Whitford have made extensive observations of this kind. Baade points out that the observed color indices are grouped around two values. The E, Sa, and Sb galaxies have colors around $+0.87$ (if we consider only the central regions of the Sb objects), about the color of a star of spectral type G. Extensive photoelectric data show that the very open Sc spirals are not so red, with colors around $+0.45$, like those of F-type stars.

These colors are strongly influenced by those of the brightest stars in each galaxy. Thus, if a system contains many red giants and relatively few main-sequence O and B stars, it will appear redder than one whose brightest members are early-type, main-sequence stars.

We can further analyze the contents of a galaxy by means of its spectrum. For all but the very nearest systems, however, the spectrum is of the integrated light — the combined radiation from all the stars and other objects in the galaxy. Morgan and Mayall have recently explored the possibility of using certain dominant spectral features as indicators of the approximate form of the spectrum–luminosity, or H–R, diagram of a galaxy's stars.

For instance, they have found that the Sb spirals (such as M 31) have absorption lines which suggest that the relative number of dwarf stars of spectral type K to giants of this type is no greater than 50 to 1. Thus the giants are relatively abundant, as they are in the neighborhood of the sun, where the ratio is 40 to 1. These K giants are so luminous that they tend to dominate the spectrum in certain regions.

Morgan and Mayall note that Sc spirals and irregular galaxies have spectra showing strong absorption lines of hydrogen. Irregular systems probably contain many main-sequence stars of types B, A, and F; their H–R diagrams might resemble that of a young star cluster. The Sc spirals also show indications of a well-populated main sequence, as is suggested by their colors. The Sb and Sa spirals, on the other hand, have spectra indicative of many cool giants and few main-sequence objects; their H–R diagrams might be like those of the older galactic clusters. Finally, the composition of the elliptical galaxies bears a strong resemblance to that of globular clusters.

These correspondences suggest that the evolution of the galaxies reflects the development of the stars they contain. Thus, elliptical galaxies would be the oldest, and the irregular galaxies the youngest, in an evolutionary sense. As with the stars themselves, however, the evolutionary age may not be the same as the chronological age.

Morgan has proceeded further and has devised a new system of classifying galaxies, in which the class, as estimated from the form of a galaxy, also indicates its spectral properties. In establishing this system, the Yerkes Observatory astronomer had the aid of the remarkable collection of galaxy photographs taken by Hubble and others with the 100- and 200-inch telescopes.

There are two extremes in the new Morgan classification. One is represented by certain spirals with little or no central condensation of luminosity, and by irregular systems like the Magellanic Clouds, for the composite spectra of these galaxies show that they are rich in stars of spectral types B, A, and F. Morgan calls these "a" galaxies. At the other extreme are the giant elliptical objects, such as those in the Virgo group and spirals like M 31. Their luminosity comes mainly from a bright central region, and they have spectra indicating that late-type stars of spectral type K are dominant. Such systems are classed as "k."

Intermediate between these extremes are af, f, fg, g, and gk galaxies, these letters being in the same order as the A, F, G, K of the sequence of stellar spectra. The principal criterion is the extent to which brightness is concentrated in the central region of the galaxy, k objects having high concentration and a ones very little. The luminosity of the brighter parts of k systems is due principally to yellow giant stars, whereas in the a galaxies the early-type stars have a pronounced effect on the spectrum in the blue and violet regions.

To the fundamental characteristics just described, Morgan appends letters for the form: S, spiral; B, barred spiral; E, elliptical; I, irregular (following the classical Hubble system for these four); Ep, elliptical with well-marked dust absorption; D, rotational symmetry without pronounced spiral or elliptical structure; L, low surface brightness; and N, systems having small brilliant nuclei superimposed on a much fainter background.

Lastly, a number from 1 to 7 can be added, 1 indicating a galaxy seen full face (at right angles to its principal plane), 7 an edgewise system (its principal plane in the line of sight). Unlike the other two parameters, this one depends solely on the angle of view, and is not an intrinsic property of the object.

For example, the Whirlpool nebula in Canes Venatici, M 51 (Figure 21), is called fS1 on the Morgan system. The older Hubble classification was Sc, that of a fairly open spiral. NGC 221, another companion of the great Andromeda galaxy is called kE3, in place of Hubble's E2. M 31 (Figure 24) itself is kS5.

Among the immediate results arising from Morgan's work is the remarkable fact that the different galaxy types are not well mixed over all parts of the sky. He wrote:

"In the area from 10 to 12 hours of right ascension and between declinations $+30°$ and $+60°$, the number of 'a' systems among the nearer galaxies is outstanding. Perhaps the most interesting comparison is between the brightest members of the Ursa Major and Virgo clouds; the former contains a large percentage of 'a' and 'f' systems, and few, if any, ellipticals; the ratio of 'a' + 'af' systems to 'k' systems is quite different in the two clusterings."

Is this nonuniformity due to chance? Or does it signify a fantastic difference in chemical composition (or other characteristics affecting evolution) of different parts of the original intergalactic material? Morgan refrains from commenting on his remarkable discovery.

IV. RADIO ASTRONOMY

Optical observations made from the surface or the earth are limited by the opacity of air. On the violet side all electromagnetic radiation is cut off at a wave length at approximately 3,000 Ångstrom units, or about 3×10^{-5} centimeter. Only by means of rockets and space vehicles can the radiation of shorter wave lengths be detected. In the red part of the spectrum air does not have a sharp cutoff, but heavy absorption bands of different kinds of molecules make it difficult to secure observations at wave lengths greater than about 10,000 Ångstroms, or about 10^{-4} centimeter. The upper atmosphere is opaque to radio waves of wave lengths shorter than approximately 1 centimeter, and to those longer than about 20 meters. Astronomers, therefore, depend mainly upon the windows, one in the optical region and the other in the radio region. The latter window is particularly useful because cosmic dust clouds cause no obscuration. Hence, a radio telescope penetrates through the opaque layer of dust and the heavy dark clouds of the Milky Way, making it possible to obtain observations of the invisible central nucleus of the galaxy and of the spiral arms that are located on the other side of the nucleus at distances of 50,000 or 60,000 light years.

The science of radio astronomy had its beginning in 1932 when Karl G. Jansky of the Bell Telephone Laboratories used a simple antenna (Figure 53) to study the direction of arrival of radio noise at high frequencies and immediately discovered that: "The first complete records obtained showed the surprising fact that the

Figure 53. K. G. Jansky's radio telescope.

azimuth of the direction of arrival of these waves changed nearly 360° in 24 hours and subsequent records showed that each day an azimuth of 0° (south) was reached approximately 4 minutes earlier than on the day before. These facts lead to the conclusion that the direction of arrival of these waves remains fixed in space, that is to say, its right ascension and declination are constant."

Jansky's first published paper on this subject appeared in the *Proceedings of the Institute of Radio Engineers* in 1933. In December of the same year he published an article in *Popular Astronomy*, in which he presented a discussion of his observations with suitable diagrams which proved that ". . . electromagnetic waves in the earth's atmosphere . . . apparently come from a direction fixed in space. The data give for the coordinates of this direction a right ascension of 18 hours and a declination of −20 degrees."

Jansky's apparatus consisted of a sensitive short-wave radio receiving system, to which he had connected an automatic signal intensity recorder. The antenna system was highly directive in the horizontal plane and was rotated continuously about a vertical axis once every 20 minutes, so that the azimuth of the direction of arrival of the radio noise could be established with a reasonable degree of accuracy, while the altitude of the signal could not be measured with precision. The wave length used was 14.6 meters, but a few additional runs were obtained at 15 and 13 meters without any appreciable difference in the resulting tracings.

Jansky quickly established that the source of the noise was somehow connected with the Milky Way, and he determined its right ascension at approximately 18 hours. Because of the changing angle that the Milky Way makes with the horizon at different times of the day and different times of the year, Jansky was able to determine an approximate value of the declination of the source. For this he gave a value of 20° south. We now know that this coincides approximately with the direction toward the center of the Milky Way. As far as I know, the only astronomers who took an active interest in Jansky's results were F. L. Whipple and J. L. Greenstein at the Harvard Observatory. In Volume 23, page 177, 1937, of the *Proceedings of the National Academy of Sciences,* they published an article in which they presented a theoretical discussion of Jansky's results, and established that the radio noise observed by Jansky could not be explained in terms of secondary solar radiations, produced in the earth's atmosphere. Their final conclusion was stated as follows: "In view of the failure of black-body radiation to account for Jansky's observations quantitatively it is particularly necessary to investigate more thoroughly the actual dependence of the received intensity on wave-length, a problem which is now being attacked by one of the authors."

The great majority of astronomers, however, took little interest in the work of Jansky, partly because they were not acquainted with the technical aspects of radio engineering and partly because they were not sufficiently receptive to the revolutionary aspects of the new discovery.

The next great step in radio astronomy was made by Grote Reber, who had read Jansky's papers and was able, as a radio engineer, to appreciate its importance. In the middle of the 1930's he built for the first time a parabolic radio antenna, which was fixed in azimuth to coincide with the meridian of his home in Wheaton, Illinois, but which could be rotated around a horizontal axis to point toward different altitudes in the sky. Reber obtained a long series of measurements that he published, partly in the *Proceedings of the Institute of Radio Engineers* and partly in the *Astrophysical Journal*. He confirmed Jansky's results at somewhat shorter wave lengths of the order of 1 or 2 meters, and constructed a map of the sky in radio waves, which bore a striking resemblance to the optical appearance of the Milky Way. In his first article in the *Astrophysical Journal*, published in June, 1940, he announced the occurrence of a maximum of intensity, coinciding with the transit of the Milky Way across the 3° acceptance cone of his antenna. He also stated that a few bright stars such as Vega, Sirius, Antares, Deneb, and the sun gave negative results.

His next paper in the *Astrophysical Journal*, in the November, 1944, issue, announced the discovery of a relatively faint radio emission from the sun, leading to the conclusion that "If . . . the Milky Way were made of average stars like the sun, a very large area in Sagittarius would have a visible intensity equal to that of the sun. Since this is not the case, some other cause must be found to make up the difference of 20 or 30 magnitudes." The fact that the sun is relatively faint in radio frequencies, compared to the background of the Milky Way, constituted one of the most sensational discoveries in astronomy during the present century. It is now well known that all of the observed radio radiation from the sky comes from diffuse nebulosities of different kinds located in the Milky Way or in other galaxies. No real star, other than the sun, has as yet been found to possess a measurable amount of radio radiation. Nevertheless, the fact that the sun is a radio source whose intensity is enormously enhanced during the outburst of a solar flare makes it reasonable to believe that many stars that resemble the sun in other physical characteristics will also be found to emit measurable amounts of radio radiation during their flare activities. Reber's radio-frequency isophotes of the Milky Way indicate not only a strong source at the center of the galaxy in Sagittarius but two or three additional sources in Cygnus, Cassiopeia, and Canis Major. Reber was thus first in having discovered discrete radio sources other than the so-called Sagittarius A source that coincides with the galactic center.

His work, more than Jansky's, made a strong impression upon the astronomers. It was frequently discussed in meetings, and led to an effort on the part of Walter Baade and Rudolph Minkowski to start some radio astronomical observations in the United States. These efforts were, however, put aside temporarily because of the outbreak of the Second World War. During the war great advances were made in the development of various electronic devices, especially in connection with radar studies, and the receivers for radio observations were greatly improved over the relatively simple devices that had been used by Jansky and Reber. The stage was thus set for the development of radio astronomical instruments, and advantage of these developments was taken first in England and Australia, and somewhat later in the Netherlands (Figure 54), France, Canada,

Figure 54. Radio telescope located at Dwingeloo, the Netherlands.

Figure 55. The 85-foot radio telescope of the National Radio Astronomy Observatory at Green Bank, West Virginia (operated by the Associated Universities, Inc., under contract with the National Science Foundation).

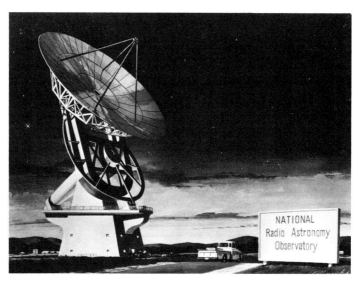

Figure 56. Artist's conception of the 140-foot radio telescope of the National Radio Astronomy Observatory (to be completed in 1963).

the Soviet Union, and the United States (Figures 55, 56). Among the thousands of radio sources often, but mistakenly, called radio stars, that have since been catalogued at the principal radio observatories, only a few per cent have been identified with objects recorded on direct photographs. These objects fall into several groups. A few are definitely known to be located within our own galaxy and are probably the remnants of former supernovae (Figures 57 to 59). Others are gaseous nebulosities, often designated as H II regions in the Milky Way, such as the nebula in Orion, or near η Carinae (Figure 59). Still other radio sources are known to be ordinary galaxies. Some of the most intense radio sources are pairs of galaxies that, from their optical appearances, are believed to be in the process of collision (Figures 60, 61). All of these identifications account for

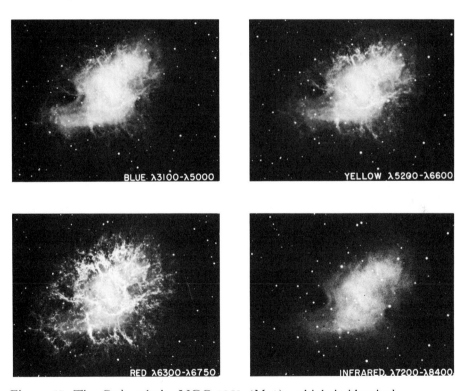

BLUE λ3100–λ5000

YELLOW λ5200–λ6600

RED λ6300–λ6750

INFRARED λ7200–λ8400

Figure 57. The Crab nebula, NGC 1952 (M 1), which is identical with the strong radio source Taurus A. The nebula is the remnant of a supernova that exploded in the Milky Way in 1054.

(Mount Wilson and Palomar Observatories photograph.)

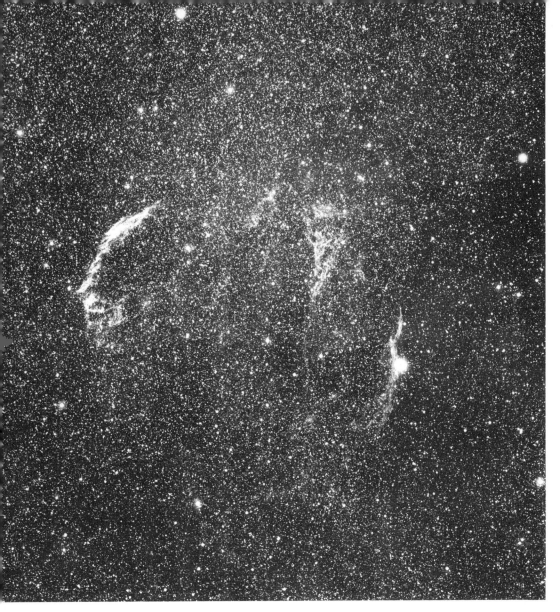

Figure 58. Loop nebula in Cygnus. Remnant of a supernova that exploded about 50,000 years ago. The nebula expands radially. It is a radio source.

(*Mount Wilson and Palomar Observatories photograph.*)

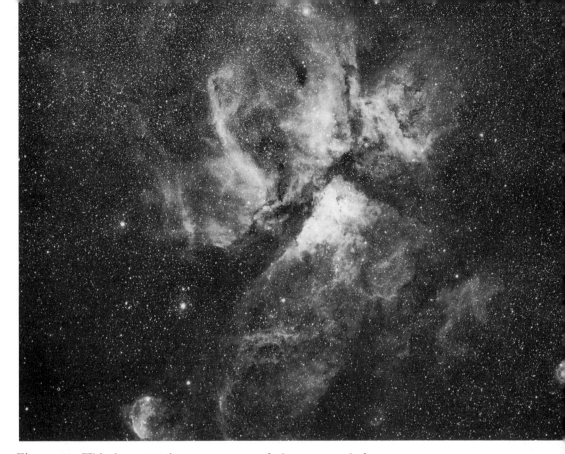

Figure 59. This is a 60-minute exposure of the great nebula near the star η Carinae, made on May 7/8, 1951, by Willem D. Victor, who used a red-sensitive emulsion with an accompanying red filter. The η Carinae nebula (right ascension 10^h 43^m, declination $-59.4°$) is one of the most beautiful bright-line emission nebulae. Its distance from the sun is approximately 4,000 light years. The diffuse nebula marks the center for an extended region of bright nebulosity and indicates in all probability the direction toward a "knot" in one of the spiral arms of our own Milky Way galaxy. This photograph shows not only the bright nebulosity and all its fine and intricate details but also the overlying network of wisps of dark nebulosity, produced by small particles that obscure the stars and emission nebulosity that lie beyond.

(*Harvard Observatory photograph.*)

Figure 60. Colliding galaxies NGC 4038 and 4039.

(Mount Wilson and Palomar Observatories photograph.)

less than about 10 per cent of the known radio sources. It is not yet known whether the unidentified radio sources are colliding galaxies, located at very great distances, and are, therefore, invisible on direct optical photographs. It is possible that some of the known radio sources may represent a type of object previously unknown to astronomers.

Figure 61. A pair of colliding galaxies known as the strong radio source Cygnus A. Its distance is about 1 billion light years. The two galaxies move with respect to each other at a velocity of 3,000 km/sec. The radio source is about 500,000 light years across, while the optical image is only 100,000 light years in extent.

(*Mount Wilson and Palomar Observatories photograph.*)

The sun is also a radio source, as are the planets, but in its quiescent state the radio emission of the sun is weak, and it could not be detected in a solar-type star located at a distance of several light years; but the radio emission of the sun varies in intensity. At certain times, especially when a strong solar flare is seen in optical light, the radio emission may be thousands of times greater than it is when the sun is quiescent. It can be shown that a solar-type star, possessing the same type of flares as does the sun, would be observable with the most powerful existing telescopes during a violent outburst even at distances of the order of approximately 10 light years.

In this connection, particular interest attaches to the so-called flare stars. Only in the neighborhood of the sun is a reasonably complete census of the stars possible. These nearby faint red dwarfs are very common and are easily found by their large proper motions. Some of them are especially interesting because they undergo sudden, short-lived increases in brightness, earning them the designation *flare stars*.

One of the first astronomers to record such a stellar flare was W. J. Luyten, using Steward Observatory photographs taken on December 7, 1948, of the proper motion star L 726-8. They revealed that in the course of a few minutes the star's brightness had risen two or three magnitudes.

Earlier that year, at Luyten's suggestion, the same star had been observed spectrographically at the Mount Wilson Observatory by A. H. Joy and M. L. Humason. The image in the focal plane of the 100-inch telescope was double, showing that the star is a close visual binary. The spectrum was typical of a dwarf M5 star, but had strong emission lines of hydrogen and ionized calcium. On September 25, 1948, however, bright lines of neutral and ionized helium appeared, and the violet-ultraviolet continuous spectrum was strengthened.

During this exposure, which lasted 144 minutes, Joy and Humason noted an increase in brightness of the system. They stated: "The change in spectrum and brightness was doubtless similar to that observed by Luyten on December 7 of the same year. Luyten suspects that the flare-up was due to line emission originating in the fainter star of the pair; our evidence indicates that the increased brightness arose mostly from the continuous spectrum."

The fainter component of L 726-8, now known as the flare star UV Ceti, is the prototype of this class of variables, though previously A. van Maanen had reported two other red dwarfs that undergo transitory increases in brightness. These stars, WX Ursae Majoris and Ross 882, had not received, however, the attention they deserved.

Recently M. Petit compiled a list of 20 objects of the UV Ceti type (Table 2). In general, the spectra of flare stars are between dM2 and dM6, but at least one is of type K2. All normally have emission lines of hydrogen and ionized calcium. During the flare outbursts the hydrogen lines are strengthened, emission lines of neutral and ionized helium and singly ionized iron become visible, and the continuum in the ultraviolet is enhanced. All stars of this type are intrinsically faint, and are located on or near the main sequence of the Hertzsprung-Russell diagram.

Table 2. Variables of the UV Ceti Type Listed by M. Petit.*

Star Name	R.A. (1900)	Dec.	Apparent Magnitude (Photo-visual)	Change in Mag-nitude	Spectrum	Absolute Magnitude (Visual)
	h m	° ′				
Wolf 47	0 57.0	+61 50	13.7	?	M3e	13.5
UV Ceti	1 34.0	−18 28	12.9	5	M5.5e	15.9
V371 Orionis	5 28.6	+ 1 53	11.7	1.8	M3e	—
PZ Mono-cerotis	6 43.2	+ 1 13	9.5	0.7?	K2e	—
YZ Canis Minoris	7 39.5	+ 3 48	11.6	1.4	M4.5e	12.5
BD +33°1646B	8 02.6	+33 06	11	?	Me	8.6
AD Leonis	10 14.2	+20 22	9.5	0.5	M4e	11.1
Wolf 359	10 51.6	+ 7 36	13.5	1	M6e	16.5
W Ursae Majoris	11 00.5	+44 02	14.8	1.8	M5.5e	16.0
V645 Centauri	14 22.8	−62 15	11.3	1	M5e	15.5
BD +55°1823	16 14.9	+55 32	10.1	0.5	M1.5e	8.4
Ross 867	17 16.1	+26 36	13.4	1	M5e	13.4
BD +51°2402	18 31.6	+51 39	8.3	?	K6e	8.0
V1216 Sagit-tarii	18 43.6	−23 57	10.5	0.4	M4.5e	13.3
W 1130	20 02.7	+54 10	12.2	?	M3e	11.0
BD −32°16135	20 35.6	−32 47	12.5	?	M3e	12.4
DO Cephei	22 24.4	+57 12	11.4	1.5	M4.5e	13.3
BD −21°6267B	22 33.3	−21 08	10.2	?	M4.5e	—
EV Lacertae	22 42.6	+43 49	10.2	2	M4.5e	11.7
EQ Pegasi	23 26.7	+19 23	11.3	0.35	M4e	11.1

* *Étoiles à sursauts lumineux*, Colloquii del Centro di Astrofisica del C.N.R., No. 2, Paris, 1958.

Flares occur at irregular intervals, and their amplitudes vary greatly in any one star. For instance, the outbursts of UV Ceti are usually about one or two magnitudes; but in September, 1952, two outbursts separated by one week had amplitudes of 3.4 and about six magnitudes. In the former the decline to normal brightness lasted only 8 minutes, but in the latter it took nearly 2 hours.

Although the outbursts are not periodic, each flare star has its own characteristic mean interval, amounting to around 1.5 days for UV Ceti. Between the flares of this star, irregular but continuous variations with a total amplitude smaller than 0.8 magnitude and a mean interval of about 30 minutes have been recorded.

One very interesting property of flares is the great rapidity of the increase in brilliance. During the outbursts of September, 1952, UV Ceti brightened by about a quarter of a magnitude per second. This was 500 times faster than the rise in brightness of Nova Aquilae 1918.

There are some indications that the flare frequency of a particular star changes slowly with time. V. Oskanjan reported at the 1956 Burakan symposium on nonstable stars that no flares of UV Ceti were seen in the observing seasons 1949–50, 1953–54, and 1954–55, but that they were numerous from 1950 to 1953. As Petit remarked, "the number and the violence of the outbursts . . . are, like the chromospheric eruptions of the sun, subject to a cycle whose duration is a property of each star." Unfortunately, we do not have enough observations to establish flare-star activity cycles resembling the 11-year cycle of the sun's activity, though 6 or more years has been suggested for UV Ceti.

Since the outbursts are easily observable, and the stars listed in the table are relatively bright, amateur astronomers could make a very important contribution to astrophysics by providing the large amount of observational data needed to determine these periodicities.

There is very little doubt that stellar flares are physically the same as those of the sun. From the spectra of flare stars, several investigators have estimated the temperatures of the disturbed regions to be of the order of 10,000° absolute. The normal undisturbed surface of a flare star has a temperature of 3,000°. With Planck's law of radiation, we can calculate that the visual surface brightnesses of the disturbed and undisturbed regions are in the ratio of 500 to 1.

In an ordinary stellar flare, where the brightness rises by about 1.5 magnitudes, the integrated luminosity of the star has increased only by a factor of 4. Therefore, T. Walraven concluded that during such a flare the area of the star affected by the disturbance represents only $\frac{4}{500}$, or about 1 per cent of the entire surface.

Of course the temperature of a very violent flare may easily be much higher than 10,000°, and the ratio of surface brightnesses may be greater than 500. Moreover, the integrated luminosity may increase by more than four times. For example, the 1952 increase of six magnitudes in the brightness of UV Ceti corresponded to a factor of 250. But, even for this violent flare, the affected area may have been only a small fraction of the star's surface.

Flares on the sun are often very strong emitters of long-wave radio energy. Hence, with a radio telescope tuned to a particular frequency, we can measure the intensity of radiation coming from a disturbed solar patch. Knowing this intensity, we can assign a "radiation temperature" to the flare region. This is the temperature that a perfect radiator (black body), similar in dimensions to the disturbed area, would need in order to radiate the same intensity at the frequency at which the measurements were made.

The radiation temperature is usually not the real temperature of a disturbed patch. The two would be the same only if such an area were optically very thick and isothermal — of uniform temperature. In reality, solar flares are optically thin. Nevertheless, the concept of radiation temperature is useful in many applications.

During the great solar flare eruption of March 8, 1947, the radiation temperature of the disturbance was reported to have been greater than $10^{15}°$. If we assume that $10^{16}°$ was reached at a frequency of 60 megacycles per second, corresponding to a wave length of 5 meters, then Planck's radiation law indicates that the total amount of energy received on the earth from this active region was 10^{-12} watt per square meter per cycle per second.

But the resolving power of our best radio telescopes is too small to observe only the disturbed area — what we measure is the total energy from the entire sun. Optical observations suggest that during a violent solar flare only about 1 per cent of the solar disk brightens, while 99 per cent remains quiescent. We conclude that the radiation from the entire quiescent sun at 5 meters was of the order of $\frac{1}{100}$ the value of 10^{-12} observed during the flare, or 10^{-14}.

How does this compare with the ordinary thermal emission of a black body having a temperature of $6,000°$ — the derived temperature of the sun's photosphere? We again apply Planck's law and find that at a wave length of 5 meters the solar radiation would be roughly 10^{-18}, or 10^{-19} watt per square meter per cycle per second. Thus, even though only 1 per cent of the solar surface is affected during a flare, the 5-meter radio emission is increased to more than 10,000 times that of the photosphere alone.

Suppose that the flares of UV Ceti resemble the sun's and occupy 1 per cent of the star's surface. The distance of the sun from us is 1.5×10^8 kilometers, while UV Ceti is 2.6 parsecs, or 8×10^{13} kilometers away. Because of the inverse-square law for the propagation of radiation, the observed emission from UV Ceti during a violent flare should be about 3×10^{11} times weaker than that of the disturbed sun, or 3×10^{-26} watt per square meter per cycle per second at 5 meters. According to I. S. Shklovsky, this amount of radiation could be detected with existing radio telescopes.

Similar considerations, of course, apply to other nearby flare stars. While we have assumed that stellar flares closely resemble those actually observed on the sun, at times the stellar outbursts are probably far more violent. Thus, there is every reason to search for radio emission among the nearer flare stars.

As far as I know, few such observations have been made. The difficulty is that really violent outbursts are rare as well as unpredictable, even in UV Ceti. But perhaps with a larger radio telescope, like the 140-foot paraboloidal antenna now under construction for the National Radio Astronomy Observatory, some of the less intense stellar flares will be recorded.

All of these ideas have been repeatedly advanced by A. Unsöld, whose discussion in the 1955 edition of *Physik der Sternatmosphären* should be read by every radio astronomer. It may even be that he was correct in his attempt to explain a large fraction of the diffuse radio radiation of the Milky Way galaxy in terms of the integrated effect of a very large number of radio stars, many supposedly in the flare stage at any given time.

His proposal has, however, been severely criticized by Shklovsky, who argues that, no matter how we modify the model of stellar distribution of the galaxy, we would require many more radio stars than the total number of all stars in order to account for the diffuse radio radiation. R. Hanbury Brown believes that most of this radiation may come from peculiar galaxies that are photographically faint. But he considers that some of the diffuse radiation may actually originate in a large number of real radio stars.

As I have said, no one has as yet succeeded in detecting the radio emission of a flare star during its outburst. One reason for this may be the fact that the flares occur at irregular intervals, that in any one flare star a violent outburst may occur once in several years, and that the duration of such an outburst may be of the order of only a few minutes and probably never longer than 1 or 2 hours. It is, however, only a question of time before we shall be able to observe real stars by means of radio telescopes. It is also reasonable to believe that not only the so-called flare stars will become observable but also many other stars of solar type which resemble the sun in temperature, intrinsic luminosity, and radius. It is even possible that radio outbursts may become observable in all types of stars and that the only reason they have not been detected as flare stars by optical means consists in the strong optical intensity of their continuous spectrum in the undisturbed regions. If the sun were 10 light years away, we should not be able to observe its flares in the integrated optical light. Nevertheless, the radio emission of an invisible flare could easily be sufficiently powerful to make it apparent in the centimeter or meter range of wave lengths.

It is important to recognize that at the present time there is little hope that we shall be able, in the foreseeable future, to observe the radio emission of stars in their quiescent stages. Moreover, because of the low resolution of a radio telescope, it would always be difficult to identify with certainty a faint but constant radio source with any given star. Since the entire background of the celestial sphere is covered with thousands of radio sources that are not of stellar origin, there would always remain some uncertainty as to whether any constant source of radio emission comes from a star or the background. The discovery of radio stars would probably depend upon the variability of the source, as indicated by the unexpected appearance of flares.

It is easy to speculate upon the advance in knowledge that will come from the discovery of true radio stars. If the flares of other stars resemble those of the sun, they will be found to be of nonthermal origin. The intensity of the radio radiation would be found to increase with the wave length, at least within the observable part of the radio spectrum. We shall undoubtedly find that in any given star the frequency of outbursts is of a periodic character, resembling the 11-year period of activity of the sun. We shall probably find that many stars possess flares of greater violence than does the sun, while in other stars the flares may be less conspicuous. There is no doubt that the study of radio stars will give us information concerning the physics of the stars, analogous to what has been discovered from the study of stellar spectra and stellar brightnesses in the relatively small optical window. I believe that at the present time the only deterrent is the relatively large amount of observing time that would have to be devoted to a single star in order to observe it during a flare, but the number of radio telescopes is increasing rapidly, and there are impressive improvements in the development of sensitive receivers and large collecting arrays. In a few years it will be possible to devote a substantial amount of time to carrying out the necessary observations.

The observations that will probably lead to the discovery of real radio stars will be made in the continuous spectrum. However, during the past few years great advances have been made in the study of the structure of the galaxy from observations of the discrete hydrogen line at a wave length of 21 centimeters. This discrete radiation appears in emission when the hydrogen gas is relatively cold and, therefore un-ionized. When a cloud of cold hydrogen is projected in front of a source emitting a strong continuous radio spectrum, the 21-cm line appears in absorption. Most of the results that have been obtained to date on the distribution of interstellar hydrogen throughout the Milky Way have come from the Leiden Observatory and the Radio Physics Laboratory in Sidney, Australia.

The cold hydrogen gas is strongly concentrated toward the central plane of the Milky Way. On the average, at a distance of about 300 light years above and below the plane, the density of the gas is about one-half of what it is in the plane itself. The thickness of the layer is, therefore, about $\frac{1}{100}$ of the diameter of the disk. The gas is ionized only in the vicinity of hot stars. The Dutch astronomers estimate that the ionized part is of the order of 5 per cent of the total amount of hydrogen. Observations of exterior galaxies have shown that the gas is not entirely confined to the central plane but is distributed in a roughly spherical halo concentric with the nucleus of the galaxy. In some cases the hydrogen gas was found to extend far beyond the optically detectable features of the galaxies. This is particularly true in the case of pairs of colliding galaxies. R. Minkowski has called attention to the work of J. Bolton on the radio source, Centaurus A, whose identification with the optical galaxy NGC 5128 cannot be doubted. The optical image of this object has a major axis of about ½ °. Its distance may be of the order of 20 million light years. The radio source associated with this galaxy consists of symmetrically spaced condensations that are separated by about 4°, or more than 1 million light years. The total extent of the radio source is almost 10°, or approximately 3 million light years. It is quite clear that the dimensions of some peculiar galaxies, when measured in the radio range of wave lengths, exceeds by a very large factor the size of the optical image. There is as yet no satisfactory explanation of this phenomenon, but it is undoubtedly related to the properties of the magnetic fields of the galaxies, which may cause concentrations of particles at great distances from the nuclei of the objects. In the case of our own galaxy, the exterior spherical halo of gas has not yet been observed. Its density can, however, be estimated by an indirect procedure to be of the order of 1 atom per 100 cubic centimeters, and its temperature may be as high as 1 million degrees. The gaseous constituents of the Milky Way in and near the central plane are known from optical observations to participate in the phenomenon of galactic rotation, discovered by Oort in 1927. The radio observations of the 21-cm line make it possible to investigate the circular velocity of the cold hydrogen in many spiral arms of the Milky Way. In the vicinity of the sun this velocity is of the order of 200 or 300 km/sec. At distances greater than 30,000 light years from the galactic center the circular velocity slowly decreases, in accordance with Kepler's third law. As we approach the galactic center, the circular velocity slowly decreases, reaching a value of 180 km/sec at a distance of about 15,000 light years from the center. At distances between about 10,000 and 1,000 light years from the center, the circular velocity again increases, reaching a value of the order of 260 km/sec. However, Oort and Rougoor have recently found a prominent spiral arm at a distance of about 10,000 light years from the center, which not only participates in the phenomenon of galactic rotation but appears to expand in the central plane, away from the galactic center. The velocity of this expansion, as measured by the Doppler shift of the hydrogen absorption line seen against the continuous spectrum of the galactic nucleus, is approximately 50 km/sec. Those parts of the spiral arm, at a distance of about 10,000 light years from the center, that are more remote than is the center itself show no absorption because their radiation does not appear superposed over the background continuous spectrum of a more distant source. However, the hydrogen emission line from this more remote part of the spiral arm produces a Doppler shift that can also be interpreted as a combination of galactic rotation and radial expansion.

The expansion of the gas poses an interesting problem. According to Oort and Rougoor, the central disk of the Milky Way would be depleted of its hydrogen in about 10^7 or 10^8 years. Since this interval is short, compared with the age of the Milky Way, and since we do in fact observe a heavy concentration of hydrogen in the central nucleus, there must be a mechanism at work which continuously replenishes the hydrogen that escapes within the central plane. The Dutch astronomers consider it most probable that the replenishment occurs as the result of the inflow of gas into the nucleus at its poles from the galactic halo. A tentative sketch of the distribution and motions of the inner parts of the Milky Way, according to Oort and Rougoor, is shown in Figure 62.

The resolving power of the Dutch radio telescope at the wave length of 21 centimeters is about 0.4° or 0.5°. The 85-foot radio telescope at Green Bank has about the same resolution. At the distance of the galactic center from us (30,000 light years) this corresponds to about 200 light years. The small black spot in the figure is about of this size: it cannot be further resolved with the existing telescopes working at 21 centimeters. This small central body is surrounded by a larger cloud of hydrogen whose radius is approximately 1,000 light years; and this is, in turn, surrounded by a ring-like feature whose radius is about 2,000 light years. Oort and Rougoor describe their diagram as a "sketch of the possible distribution of hydrogen in the central part of the galactic system." Nevertheless, the general features of the structure of the radio source "Sagittarius A" are probably correct. The density of H in the ring is about 1 atom per cm³. There "appears to be a practically empty space within the ring," but at a distance of about 1,000 light years from the center the density of H is again of the same order of magnitude. At a distance of 300 light years from the center it is about 3 atoms per cm³. Still closer to the center "the density must increase very strongly."

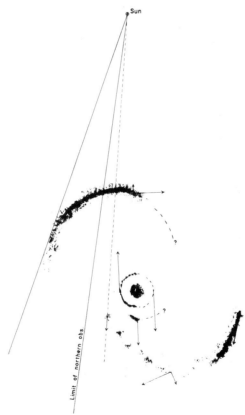

Figure 62. Sketch of possible distribution of hydrogen in the central part of the Milky Way according to G. W. Rougoor and J. H. Oort.

In order to obtain better resolution, F. D. Drake has observed the galactic center with the 85-foot radio telescope at Green Bank, using a receiver tuned to a wave length of 3.75 centimeters. Since there is no discrete hydrogen line at this wave length, his observations recorded the continuous emission of the gas in the vicinity of the center. No velocity measurements could be made, but the beam width was only 7 minutes of arc, or about 60 light years.

The results are shown in Figure 63. The radio source Sagittarius A is shown to consist of two pairs of condensations, symmetrically spaced with respect to the geometrical center. The two inner condensations are roughly 40 light years from this center and are about 30 light years in radius. The outer two condensations are about ±300 light years from the center. Drake wrote: ". . . from the symmetry in the outer sources, it is suggested that they actually rep-

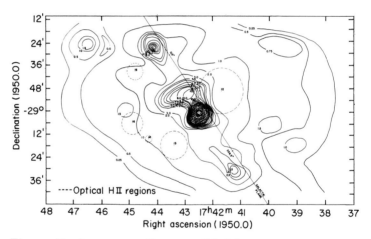

Figure 63. A contour diagram of brightness temperature at 8,000 megacycles in the region of the galactic center, made by F. D. Drake with the National Radio Astronomy Observatory 85-foot telescope. The two sources at the center, and the two in the plane about ¾° either side of the center, are the most prominent features. The other weak sources that appear out of the plane are probably H II regions.

(*Annual Report of the National Radio Astronomy Observatory for 1958–1959*, Figure 5.)

resent the projection of a ring of emission surrounding the galactic center." In drawing the isophotes of his figure, Drake has made allowance for the smoothing effect of the 7-minute beam width of the telescope.

Both pairs of sources in Drake's figure fall within the central spot of Figure 63. Although it is not possible to observe these four sources at wave lengths greater than about 4 centimeters, Drake has analyzed observations made by other workers at longer wave lengths, by assuming that the location of the four sources is known. In this manner he has determined the flux densities of the four sources at different wave lengths, and has found that the radio intensity of the two inner sources decreases with increasing wave length, in accordance with Planck's law. Accordingly, he concluded that the two inner sources have thermal spectra, the hydrogen gas being ionized as in a luminous hydrogen nebula. The two other sources show increasing radio intensities with increasing wave lengths – a phenomenon that violates Planck's law and implies that the radiation is mostly nonthermal in character, resembling the radiation observed in the Crab nebula, where it has been attributed to the acceleration of very fast electrons in a strong magnetic field. Drake, therefore, believes that the nonthermal radiation of the outer sources has something to do with the shearing effect that must be present in the outermost strata of the central hydrogen cloud which, according to the Dutch astronomers, rotates as a solid body.

The ionization of the two central sources implies the presence of hot stars whose thermal radiation is sufficient to produce the ionization of hydrogen. A very few hot O-type stars in the central body would be sufficient to produce the ionization of the medium whose density is of the order of 100 atoms per cubic centimeter. It is, however, much more probable that the central body contains stars of Baade's Population II, and Drake has estimated that their number may be of the order of 1 billion.

We have already seen that the Andromeda galaxy (M 31) is in many respects quite similar to the Milky Way. It is, therefore, reasonable to compare the structure of the nucleus of M 31 with the nucleus of the Milky Way. The photographic resolving power of a very large optical telescope is of the order of 1 second of arc. The distance of the Andromeda galaxy is about 1.5 million light years. Hence the optical resolution at the distance of the Andromeda galaxy is approximately 10 light years. Visual observations of M 31 made at Mount Wilson, Palomar, and the Lick Observatory have shown that the central nucleus of M 31 is almost star-like and that its diameter is of the order of 1 second of arc. This star-like center is surrounded by a fainter, luminous disk whose radius is about 20 light years.

A. Lallemand, M. Duchesne, and M. Walker have recently obtained spectrograms of the nucleus of M 31 with an image converter at the focal plane of the Coudé spectrograph of the 120-inch Lick reflector. The spectrum is continuous and shows strong absorption lines of ionized calcium. The light that is optically recorded, therefore, comes mostly from stars of intermediate spectral types, resembling the sun, and it is therefore reasonable to conclude, with the Lick observers, that the central body of the Andromeda galaxy contains approximately 100 million solar-type stars within a spherical volume whose radius is about 20 light years. This is not unreasonable in view of the fact that Drake had suggested 1 billion within a much larger central volume of the Milky Way.

The most startling discovery made by the Lick observers is the tilting of the absorption lines of ionized calcium in the spectrum of the nucleus of M 31 (Figure 64). The observers interpret this as a solid-body rotation of the nucleus of M 31.

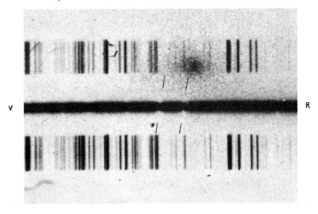

Figure 64. Spectrum of the central nucleus of the Andromeda galaxy. The spectrum is continuous and shows two inclined absorption lines of ionized calcium. *(Lick Observatory photograph.)*

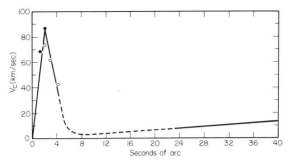

Figure 65. Circular velocity of rotation of the innermost parts of the Andromeda galaxy (M 31).

Figure 65 shows the circular velocity of rotation, which reaches a maximum of nearly 90 km/sec, at a distance of 1 second of arc, or about 10 light years from the center. At distances greater than 1 second of arc, the circular velocity of rotation in M 31 decreases, reaching a value of approximately zero, 8 seconds away from the center; it then slowly increases and at about 24 seconds of arc from the center connects with the rotational curve previously established by H. W. Babcock (Figure 66). The maximum of 100 km/sec, found by Babcock at a distance of about 5 minutes of arc, is followed by another minimum, then by a gradual increase that continues to a distance of 100 minutes of arc from the center. At still greater distances, the circular velocity again decreases, in accordance with Kepler's third law. The Lick observations, therefore, show that the very small central superglobular cluster in M 31, which rotates as a solid body with a velocity of about 90 km/sec at its edge, is surrounded by a much larger disk, which also rotates as a solid body

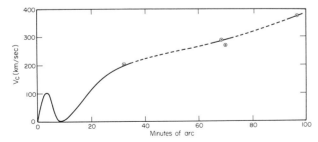

Figure 66. Circular velocity of rotation of the outer parts of the Andromeda galaxy (M 31).

with a circular velocity of 100 km/sec, and this disk gradually merges with the spiral structure of the galaxy in which there occurs a gradual transition from solid-body rotation to a Keplerian type of motion.

The resemblance between the central body of M 31 and that of the Milky Way is still further enhanced by G. Münch's recent discovery of turbulence and expansion within a distance of about 600 light years from the center of M 31. These observations were made with the 200-inch telescope on Mount Palomar and are indicated by the wavy appearance of the emission line of ionized oxygen 3726-28. When the slit of the spectrograph was oriented along the minor axis of the projected elliptical disk of the Andromeda galaxy, the tilting of the emission line could not be explained in terms of galactic rotation but indicated a velocity of expansion of the order of 50 km/sec at a distance of 600 light years from the center.

The only conspicuous difference between the center of the Milky Way and that of M 31 is the presence of two symmetrical condensations at the center of the Milky Way, while in the case of M 31 the optical data have shown only one central body.

V. • BINARY STARS AND VARIABLES

In 1834 F. W. Bessel noticed that the apparent motion of the brightest star in the sky, Sirius, is not uniform along a straight line as it would be if no force were acting upon it. When the measured positions of Sirius were plotted on a map, they described a wavy trajectory on the celestial sphere, which led Bessel to write to A. von Humboldt: "I adhere to the conviction that Procyon and Sirius are genuine binary systems, each consisting of a visible and an invisible star. We have no reason to suppose that luminosity is a necessary property of cosmical bodies. The visibility of countless stars is no argument against the invisibility of countless others."

In 1851 C. A. F. Peters re-examined the problem and confirmed Bessel's conclusion; ten years later T. H. Safford predicted that the companion of Sirius would be found in a position angle of 84° in the year 1862. An even more accurate prediction was made by A. Auwers, and in 1862 Alvan G. Clark, while testing a newly made telescope objective, discovered a faintly luminous companion approximately in the position predicted by Safford and Auwers.

Since that time the companion of Sirius has been observed by dozens of astronomers, and the orbit of the double star is now accurately known. Its period is 50 years and the semimajor axis of its elliptical orbit around the primary star is 7.6 seconds of arc. Since the distance of Sirius has been determined by means of trigonometric measurements and is now known to be 8.7 light years, the angular semimajor axis of the orbit can be expressed in kilometers or astronomical units. The result of the computations is approximately 20 astronomical units. When the period and the semimajor axis are both known, we may apply Kepler's third law to compute the total mass of the double star system. This law is given by the expression

$$(m_A + m_B)P^2 = a^3$$

in which the period P is expressed in years, the semimajor axis a in astronomical units, m_A and m_B are the masses of the bright companion of Sirius and that of its faint companion. Inserting 50 for P and 20 for a, we find that m_A plus m_B is equal to 3.2 solar masses (Figure 67).

Kepler's third law does not give us the masses of the two stars separately but only the sum of their masses. However, in the case of Sirius, the ratio of the masses of the two stars has been determined

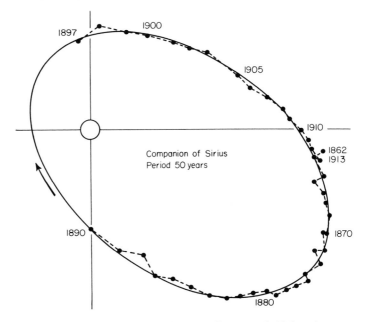

Figure 67. Relative orbit of Sirius B around Sirius A, as seen projected against the celestial sphere.

(*Courtesy of Yerkes Observatory.*)

from the orbital motion of each component. The result is that about two-thirds of the total mass belongs to the bright star and one-third to the faint companion. The former is thus a fairly normal main-sequence star having a surface temperature of about 10,000° and a spectral type designated as A1. The apparent magnitude of Sirius A is −1.4. The apparent magnitude of the companion is, however, +8.7; yet the two stars have approximately the same spectrum, the same color, and the same surface temperature. Hence, the only way we can account for the apparent faintness of Sirius B is to assign to it a much smaller surface area than to Sirius A: since the temperatures of the two stars are alike, every square centimeter on the surface of Sirius A radiates into space the same amount of light and heat as does every square centimeter on the surface of Sirius B. Since Sirius B is about ten magnitudes fainter than Sirius A, we conclude that the ratio of the luminosities of the two stars corresponds to a factor of about 10,000, and this is also the ratio of their surface areas. The surface area of a sphere is proportional to the square of the radius; hence, the radius of Sirius B is only about 1⁄100 of the radius of Sirius A. The latter is about twice as large as is the sun, so that the radius of Sirius B is roughly 1⁄50 of the radius of the sun.

But its mass is approximately the same as that of the sun, 2×10^{33} grams. Hence, the mean density of Sirius B, namely, its mass divided by its volume, is about 125,000 times greater than the mean density of the sun, which was previously found to be 1.4 grams per cubic centimeter. The gases that constitute Sirius B are thus compressed to such an extent that 1 cubic centimeter weighs 200,000 grams. A star of this type is designated as a white dwarf, and Sirius B is one of these remarkable objects. It is, however, by no means the most remarkable among them. A faint star designated by its catalogue number AC +70°8247 is believed to have a radius of only 0.005 times the radius of the sun. Its volume is, therefore, 10 million times smaller than that of the sun, but its mass is approximately equal to that of the sun; hence, the mean density of this star is roughly 10 million times greater than the density of water.

Using slightly different values, W. S. Krogdahl concluded that "at the surface of the earth a little 0.1-inch cube of average stuff from this star would weigh almost 1300 pounds. At the surface of the star, however, this same pinch of material would weigh 2,200,000 tons, for the surface gravity of this white dwarf is calculated to be 3,400,000 times the earth's. Here a 150-pound man, though his body were as strong as steel, would be crushed to tissue thinness by his own 250,000 tons of weight. The enormous compressive force of the star's tremendous surface gravity causes its atmosphere to increase in density about a billion times through a depth of just 15 feet; consequently, the whole atmosphere of this white dwarf is probably only 10 to 15 feet thick!"

The large mean densities of the white dwarfs suggest that they do not obey the laws of perfect gases. If we should disregard this point and compute the central temperatures based on the perfect gas laws, we would obtain approximately 1 billion degrees. With such an immense temperature the nuclear processes would be extremely active. If hydrogen were present, the star would blow up like a hydrogen bomb. But even without hydrogen there would be nuclear processes building up heavier atoms from helium, and these would supply far more radiation than is actually observed in the feeble glow of Sirius B. Its central temperature cannot be anything like 1 billion degrees.

In trying to find the nature of the departure from the law of perfect gases, we are at first led to follow intuition and attribute it to the van der Waals forces, which arise when laboratory gases are compressed to about 100 or 1,000 atmospheres. In this state of compression the atoms, whose diameters are of the order of 10^{-8} centimeter, begin to touch one another. But at the densities of 10^5 to 10^8 that of water, which prevail in white dwarfs, the atoms are crushed "like eggs packed in the bottom of a heavily laden basket," to use G. Gamow's apt expression.

This crushing removes many of the outer electrons from the atoms in the white dwarf's interior, and the atoms become ionized. The remaining fragments of atoms occupy individual spheres only of about radius 10^{-12} or 10^{-13} centimeter. This kind of gas can be compressed to a density billions of times greater than that of water without encountering any difficulty from the actual contact between the atomic fragments. Despite its high density, the substance of a white dwarf is gaseous, and we need not even invoke the high temperature of its interior to account for the necessary degree of ionization to avoid the effect of van der Waals forces.

Why then does the ordinary perfect gas law not apply? The answer is found in the theory of degenerate matter formulated by E. Fermi and P. Dirac, and applied to white dwarfs by R. H. Fowler and S. Chandrasekhar. The properties of a degenerate gas differ drastically from those of a normal gas. Of course no one has ever obtained such a gas in the laboratory, and our conclusions rest upon theory and such indirect confirmations as may be obtained from an analogous consideration of the problem of the solid state of matter.

Ordinarily, the pressure is proportional to the product of density and temperature. When the temperature is zero absolute, the pressure is also zero; this can be visualized by imagining that the normal full-sized atoms are brought to rest. But in the interior of the atoms the orbital motions of the ring of electrons are not brought to rest as the temperature drops to zero; even in a completely cold hydrogen atom the electron continues revolving around the proton. Nevertheless, the atom as a whole is at rest with respect to its neighbors.

In the crushed state of matter, however, the free electrons, squeezed off their original atomic nuclei, do not come to rest at zero temperature; they retain vestiges of what were once their orbital velocities around the nuclei. Thus the motions of electrons that were once orbital, and as such exerted no pressure upon the walls of an enclosure, are now more or less at random, and even at absolute zero they contribute an appreciable *zero-point pressure*.

Suppose that we start with a vessel containing a small amount of degenerate gas. From the foregoing discussion we know that even at zero temperature the gas exerts a certain pressure. We shall now add more degenerate gas to the vessel, all at zero temperature. The density will, of course, increase. But will the pressure also change? In a perfect gas at zero temperature it would always remain equal to zero.

By adding more gas to the vessel we, of necessity, increase the pressure. The argument rests upon Pauli's exclusion principle, which states that the newly added electrons cannot duplicate the velocities and spatial locations of electrons already present in the gas. Thus, if the old electrons have the smallest velocities consistent with their original orbital motions, then the new electrons must, on the average, move faster. They increase the zero-point pressure, which is proportional to the $\frac{5}{3}$ power of the density and is independent of the temperature. Only for the greatest densities found in the white dwarfs is there a modification in this law; according to Chandrasekhar, the pressure then becomes proportional to the $\frac{4}{3}$ power of the density. But we shall not use this elaboration even though it is essential in an accurate computation.

A remarkable property of degenerate stars can be deduced immediately. Consider Sirius B, whose mass is equal to that of the sun, and whose mean density is 200,000 times that of water. The mean pressure is given by the law that we have just stated: $10^{13} \times$ 200,000$^{\frac{5}{3}}$ dynes, if the gas is hydrogen (the quantity 10^{13} has been computed from quantum theory). It is this internal pressure of the gas that prevents the star from collapsing under its own tremendous gravity, and the size of the star is determined by the balancing of the pressure against the gravity.

To understand the consequences of this situation, suppose that we pack an additional amount of material, also equal to the sun's mass, into the volume occupied by Sirius B. The density will become twice what it was before. Every cubic centimeter within the star will have twice its former mass, and by Newton's law of universal gravitation will attract every other cubic centimeter with a force that is proportional to the product of their masses; hence, the attraction of the hypothetical star upon itself will be 2×2, or four times as great as in Sirius B.

But the internal pressure will also rise; by the foregoing density law it will be $2^{\frac{5}{3}} = 3.2$ times as great as in Sirius B. But this is insufficient to hold the star to its original size because of the four-times increase in internal gravity, and the star of twice the sun's mass must, therefore, contract, until it again becomes small enough to establish equilibrium between pressure and gravitation.

The remarkable property we have deduced is that the size of a degenerate star depends inversely upon its mass. In other words, the heavy white dwarfs are smaller than less heavy ones. Chandrasekhar has computed that a white dwarf that consists of pure hydrogen would have a radius of zero were its mass about five suns — there can be no degenerate star of larger mass. A white dwarf, consisting entirely of helium, cannot have a mass in excess of $1\frac{1}{2}$ times that of the sun. This results from the fact that the pressure of a crushed assembly of atoms is given by the number of free electrons. Hydrogen furnishes only one such electron per proton of unit mass, whereas helium has two per alpha particle of mass 4. But the attraction of the star upon itself depends almost entirely upon the atomic nuclei. Therefore, the balance of the two forces, gravitation and pressure, demands a greater reduction in radius in the case of helium than for the lighter hydrogen.

Mestel has considered the energy sources of white dwarfs and the possible results of their accretion of interstellar matter. If we assume that a thin outer skin of a white dwarf consists of ordinary nondegenerate matter, we can compute the pressure and temperature, step by step inward from the surface, using the ordinary laws of perfect gases. At the base of this outside layer the pressure is 10^{13} atmospheres, the density about 2 kilograms per cubic centimeter, and the temperature about 20 million degrees. Inside this layer the density becomes so great that the gas is degenerate because the atoms are crushed.

The thin outer layer contains about $\frac{1}{400}$ of the total mass of the star, but it forms a blanket of high opacity which limits the rate of cooling of the degenerate core. This layer is subject to the star's high surface gravity, which greatly broadens the lines in its spectrum; thus, the spectrum gives an important clue to the high density of the star's interior even though it arises only in the surface layer.

The temperature inside all but the outer $\frac{1}{4}$ per cent of the mass differs very little from 20 million degrees. Yet this temperature is ample to convert hydrogen into helium — if hydrogen is present. But so long as nuclear processes are active, a star cannot condense to the degenerate state; it must remain similar in constitution to the sun. Mestel concludes that the white dwarfs contain no nuclear energy sources in their interiors and, hence, no internal hydrogen, in agreement with Schatzman.

On the other hand, we know that the interstellar clouds consist mostly of hydrogen. Therefore, these peculiar stars cannot have condensed directly out of the interstellar gas into the white-dwarf state; they must have been originally ordinary hydrogen-rich stars (on the assumption that all stars are formed from interstellar matter).

Since the present mass of Sirius B is the same as that of the sun, and since nuclear evolution causes only a negligible decrease of mass, by 0.7 per cent, we might think of Sirius B as the kind of star that the sun will become after it has converted all its hydrogen into helium. This hypothesis is unsatisfactory, however, because with its slow rate of energy production the sun will require 10^{11} years to reach the stage of hydrogen exhaustion. Yet in the Hyades cluster we find four white dwarfs and several normal G and K dwarfs. There are no intermediate types. Since all the cluster members must be approximately the same age — much less than 10^{11} years — this theory is ruled out.

Mestel, therefore, adopted the hypothesis suggested by F. Hoyle and also favored previously by E. A. Milne, Chandrasekhar, Gamow, and others: What we now observe as a white dwarf was once a very massive O or B star that rapidly burned up its hydrogen and then, in the final stages of its "perfect gas" condition, blew off enough material, perhaps in a nova (Figures 68, 69) or supernova

Figure 68. Expanding nebulosity around Nova Persei.

(Mount Wilson and Palomar Observatories photograph with the 200-inch telescope.)

June 9, 1950   February 7, 1951

Figure 69. NGC 5457, spiral galaxy in Ursa Major. Messier 101. Two views with and without nova: June 9, 1950, and Feb. 7, 1951.

(Mount Wilson and Palomar Observatories photograph.)

outburst, to reach the critical mass of 1½ suns for helium, after which it could contract by its own gravitation, eventually reaching the white-dwarf state.

A white dwarf radiates energy without replenishment. It contains no hydrogen and is thus deprived of nuclear sources. It cannot contract still further and derive heat by the Helmholtz–Kelvin process because, as we have seen, its radius is permanently fixed by its present mass. It can only cool slowly, as radiation from its supply of internal heat diffuses outward through its highly opaque "perfect gas" skin at a rate which will maintain the star's present low luminosity for billions of years.

Turning now to Luyten's H–R diagram for faint stars (Figure 70): Open circles represent ordinary dwarfs (at the lower end of the main sequence), dots the white dwarfs, and square dots are the four Hyades white dwarfs. Each white dwarf is observed at only one brief moment of its existence, but all of them taken together are believed to give a picture of the evolution (or "aging") of these stars. The situation is far simpler than elsewhere in the H–R diagram because surface temperature, becoming less and less as the white dwarf cools, is the only variable factor.

Thus, the location of such a star along Luyten's sequence depends on its surface temperature and, hence, on its age: the oldest and

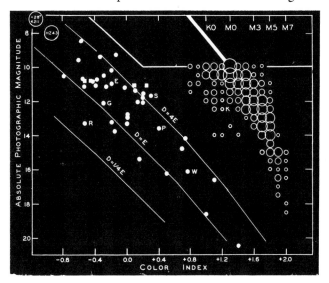

Figure 70. W. J. Luyten's diagram of stars of faint absolute luminosity. On the right are ordinary red dwarfs; to the left are white dwarfs. Stars identified by letter are the following: E, 40 Eridani B; G, AC +70° 8247; P, companion to Procyon; R, Ross 627; S, companion to Sirius; W, Wolf 489. The curves show the locations of stars with diameters four times, equal to, and one-fourth the earth's diameter, respectively.

(*University of Minnesota chart.*)

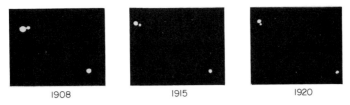

Figure 71. Three successive photographs of the visual double star Krueger 60 and of a comparison star.

(Yerkes Observatory photograph.)

reddest stars (Figure 71, for example) are on the right side of the band. Stars of greater mass have smaller radii and lie farther to the left and lower. Those along the lower edge are approximately one-half the size of the earth, and their masses are near one and one-half suns, the maximum for white dwarfs made of helium. Stars near the upper edge of the band are about four times as large as the earth, with masses about half that of the sun.

It will be noticed in Luyten's diagram that the white-dwarf sequence and the main sequence are not quite parallel. Theoretically, the luminosity of a main-sequence star is proportional to the $5\frac{1}{2}$ power of its effective surface temperature. But a white dwarf cools without change in radius, so that its luminosity will remain proportional to the fourth power of the temperature, in obedience to Stefan's law. As a result, the two sequences should have different slopes, and there is excellent agreement between observation and theory.

Since we know (or believe we know) that the more massive stars evolve faster than the less massive ones, and that the latest stage of a star's evolution is a white dwarf, we must conclude that Sirius B, which now has only one-half as much mass as Sirius A, was at some time in the past the more massive member of the binary. Sirius A is now a normal main-sequence star, and its age and, therefore, also the age of Sirius B, must be of the order of 10^7 years. Sirius B must then have had, originally, a mass of the order of 10 or more solar masses in order to have had time to run through all four stages of evolution and discard about 90 per cent of its mass before settling down as a white dwarf.

At this point in the discussion astronomers usually state that before Sirius B could have become a white dwarf it was probably a nova, and previous to that a red giant "dominating the light of the system — which must have been a few magnitudes brighter than now and distinctly reddish in color" (Z. Kopal: *Close Binary Stars*, John Wiley & Sons, Inc., New York, 1959, page 542). They then remark that the ancient Greek catalogues described Sirius as a red star, which is "quite at variance with its blue-white colour as we know it today" (*ibid.*). But they also, more or less reluctantly, conclude that Sirius B could not have evolved so drastically in about two thousand years and that "*Es ist mehr als waghalsig, wenn man wegen zweier Worte der antiken Literatur die Existenz roter Mittelsterne postulierte.*" (*Astr. Nach.*, 231, 387, 1928.) We do not know exactly in what manner massive stars get rid of their excess masses, but most astronomers believe that this may take place in the form of nova and supernova outbursts, as well as in the form of a continuous outflow of gas, as in Wolf–Rayet stars. The novae are known to occupy a region in the H–R diagram which coincides with the nearly vertical evolutionary tracks in the fourth stage of evolution. The amount of mass which is lost in a single nova outburst is quite small — about 10^{-5} of the mass of the star. But according to B. V. Kukarkin and P. P. Parenago, ordinary nova outbursts are recurring phenomena. Nova T Coronae Borealis had an outburst in 1866 and a very similar one in 1946, eighty years later. Its rise to maximum brightness, in the course of a few days, amounted to eight stellar magnitudes. Another recurrent nova, T Pyxidis, has had outbursts about every twenty years, with a range of about seven magnitudes. According to the Russian astronomers, the average intervals between outbursts in a particular star are correlated with their ranges in brightness. When a nova increases in brightness by only a magnitude or two, its outbursts occur every few years. In one particular star, 17 Leporis, the range in brightness is of the order of only about 0.1 magnitude, and its outbursts are separated by about four months. This star probably loses only about 10^{-10} of its mass in an outburst. If the relation between the time interval and range of brightness is extrapolated to ordinary novae for which Δmag is of the order of 13, the corresponding time interval should be of the order of 10^5

years. Whether the relation may be safely extrapolated to super-novae with Δmag $\sim$ 19, is not known, but as a guess we may assume that for them Δt may be as long as 10^9 years.

The range in brightness of a nova is undoubtedly a rough measure of the amount of mass lost in each explosion. For an ordinary nova this is 10^{-5} of the mass of the star. For a rapidly recurring nova it may be only 10^{-7} of the stellar mass, and for a supernova it may be 0.1 to 0.5 of the original mass.

If an ordinary nova loses 10^{-5} of its mass in a single outburst, then it would require roughly 10^4 outbursts for a fairly massive star to get rid of 90 per cent of its mass. The corresponding interval of time would be

$$10^4 \times 10^5 = 10^9 \text{ years.}$$

A supernova would require only one or two outbursts, also with a time interval of 10^9 to become a white dwarf. A nova-like object such as 17 Leporis, in which $\Delta t \sim 0.3$ year, would experience 3×10^9 outbursts in 1 billion years and lose approximately $3 \times 10^9 \times 10^{-10}$ = 0.3 of its mass before it becomes a white dwarf.

No one knows how a star of large mass "knows" that it must discard a large part of its mass before it can become a white dwarf; nor is there any satisfactory explanation for the fact that different stars lose mass in such a variety of processes. There is, however, some indication that many nova-like stars and ordinary novae are peculiar binary systems, in which one component is small and hot while the other is large and cool. In such binaries the large and cool star often expels gas that becomes trapped in the vicinity of the small and hot star. When the amount of trapped gas exceeds a certain limit, the hot star causes it to erupt, and we observe a nova.

Six years ago Merle F. Walker, then working at the Mount Wilson Observatory, made the surprising discovery that the nova DQ Herculis is an eclipsing binary star. During December, 1934, this nova reached a maximum brilliance of magnitude 1.3 but had faded to about 14 when Walker made his photoelectric observations two decades later.

There is only one other known case of a nova in a binary system. However this does not rule out the possibility that many novae may have close companions that would be very difficult to detect.

Two stars, SS Cygni and AE Aquarii, whose outbursts resemble those of ordinary novae but are less violent, are close doubles. Moreover, the period of light variation and other features of the light curve of DQ Herculis bear a strong resemblance to those of UX Ursae Majoris. Thus the duplicity of Nova Herculis may not be an isolated phenomenon.

Primary eclipses of the nova by an invisible companion recur every 4 hours and 39 minutes, but there is no indication of a secondary minimum between two primary ones (Figure 72). The latter are quite deep — about 1.2 magnitudes — which suggests that these eclipses are total or nearly so, and that the invisible occulting star has a much lower surface brightness than the nova itself.

Perhaps of greater significance is the even more remarkable discovery made by Walker immediately after he had found the binary nature of DQ Herculis. Normally, after ejecting perhaps a hundred-thousandth of its mass in a nova outburst, a star returns in the course of a few decades to its prenova state — an underluminous blue star, sometimes without any absorption or emission lines in its spectrum, and eventually shows only minor, irregular variations.

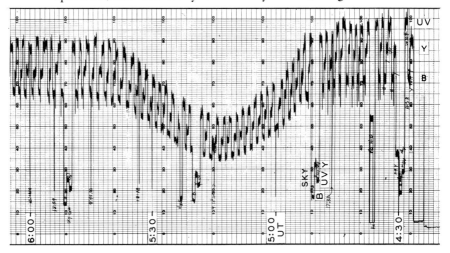

Figure 72. Part of the light curve of Nova Herculis 1934 as observed by M. F. Walker in ultraviolet (UV), yellow (Y), and blue (B) light.

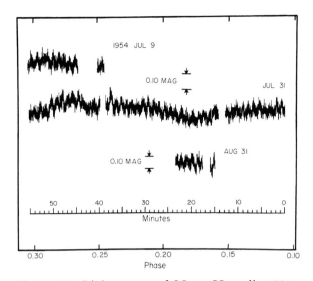

Figure 73. Light curve of Nova Herculis 1934 outside of eclipse, showing oscillations with 71-second period.

But the light curve of Nova Herculis, outside of eclipse, contains small, strictly rhythmic oscillations (Figure 73) whose period is 71.0604 seconds, with an uncertainty of only ±0.00035 second — a very striking degree of precision. The amplitude of these oscillations is not always the same; usually they amount to a few hundredths of a magnitude, but at times they are too small to be detected. However, when these waves in the light curve can be observed, the maxima and minima occur exactly on schedule. Evidently the oscillations are always present, even when unrecorded.

The strict periodicity suggests that the nova is pulsating like a Cepheid variable, but far more rapidly, since it contracts and expands in a cycle of only 71 seconds, instead of 5.37 days as for δ Cephei. As we know, there is a well-known law that holds for all pulsating stars: The period in days multiplied by the square root of the mean density (in terms of the sun's) is a constant, near 0.1 in each case. For example, the mean density of δ Cephei is 0.0006 times that of the sun; multiplying the square root of this number by the period, 5.37 days, gives 0.13 for the constant.

If we use this value in applying the law to Nova Herculis, with a pulsation period of 71 seconds or 8×10^{-4} day, we find

$$(\text{Density})^{1/2} = 0.13/(8 \times 10^{-4}).$$

Thus the mean density of the nova is about 25,000 times that of the sun, or 35,000 grams per cubic centimeter.

Although a drastic extrapolation has been made in this calculation, theoretical studies by I. Epstein in 1950 and E. Sauvinier-Goffin in 1949 lend support to it. There can be no doubt that the nova is an extremely dense star, resembling a white dwarf in this respect.

In 1956 Robert P. Kraft observed the spectrum of Nova Herculis, using the prime-focus spectrograph of the 100-inch Mount Wilson reflector. His measurements showed that the radial velocity, determined from several emission lines in the nova spectrum, varied in phase with the cycle of eclipses. Two years later a new series of spectrograms was obtained by J. L. Greenstein with the 200-inch telescope. Their papers in the *Astrophysical Journal*, the first by Kraft and Greenstein, the second by Kraft alone, summarize the results of the combined observations.

The accompanying set of 10 spectrograms supplied by Kraft and Greenstein, shows how the spectrum changes during the course of an orbital cycle of DQ Herculis. Here the phase is the fraction of a period that has elapsed since mid-eclipse, corresponding to 0.00 (Figure 74). Phases 0.25 and 0.75 represent the times of elongation of the components as seen from the earth — when the stars are side by side in the plane of the sky.

The spectrum consists of a continuous band with strong emission lines. The latter are produced partly in the large expanding nebulous envelope that was blown off during the nova's outburst in 1934, and partly in a much more compact nebula surrounding the pulsating component of the binary.

Members of the hydrogen Balmer series at longer wave lengths, $H\delta$, $H\beta$, and $H\gamma$, change little, if at all, with phase, indicating that they arise mostly in the large envelope. On the other hand, the higher Balmer lines, at shorter wave lengths, vary in intensity, being faintest during eclipse and brightest at about phase 0.6. They appear to originate mainly in the compact nebula.

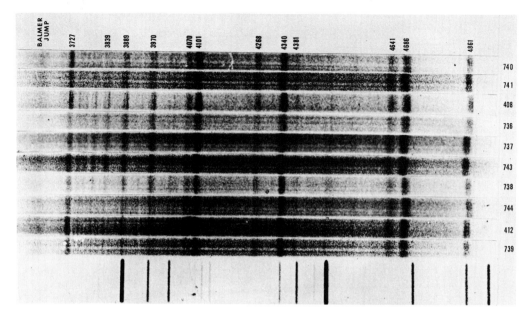

Figure 74. Ten spectrograms of DQ Herculis taken with the 200-inch telescope have been arranged in order of phase to show spectral changes during the 279-minute cycle. At the right are plate numbers; the corresponding phases, counted in fractions of the period beginning with mid-eclipse, are 0.01, 0.26, 0.32, 0.40, 0.52, 0.67, 0.67, 0.81, 0.82, and 0.82. At the top are wave lengths, in Ångstrom units, of some of the conspicuous emission lines, including 3727, a forbidden line of ionized oxygen; 3839, 3889, and 3970 — higher lines of the Balmer hydrogen series; 4070, a forbidden ionized sulfur line; 4101, Hδ; 4340, Hγ; 4686, ionized helium; and 4861, Hβ.

(The spectra are from Mount Wilson and Palomar Observatories.)

This interpretation by Kraft and Greenstein was confirmed on short-exposure spectra obtained in rapid succession during three eclipses of the nova. As the invisible companion passed in front, the higher members of the Balmer series faded much more than Hβ or Hγ (Figure 75). In the same way, it was demonstrated that the emission line of ionized helium (He II) at a wave length of 4686 Ångstroms comes mostly from the small nebula, while the forbidden oxygen line at 3727 Ångstroms originates primarily in the large envelope.

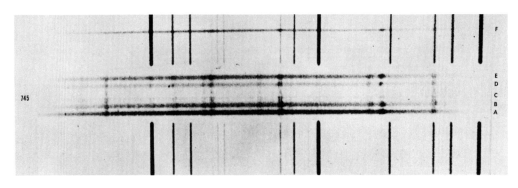

Figure 75. J. L. Greenstein took on one plate five exposures in rapid succession during a single night. Their phases, from bottom to top, are A, 0.90; B, 0.93; C, 0.00; D, 0.06; E, 0.10. The emission lines match those in Fig. 74, 3727 being at the left, 4861 at the right. Note that the helium line 4686 is much more weakened at mid-eclipse than is the oxygen line 3727, indicating that the latter originates in a large nebulous envelope surrounding the binary.

(*Mount Wilson and Palomar Observatories photograph.*)

As the illustrations show, the higher Balmer lines change in shape as well as brightness during the cycle. They are double near phases 0.25 and 0.75; in the former case the longer wave length component of each line is stronger, while the opposite is true in the latter.

The double character of Hζ at 3889 Ångstroms is easy to see in Figure 76. For members of the Balmer series toward shorter wave lengths than this, the confusion of double components among the closely spaced emission features is more difficult to disentangle. The helium line at 4686 Ångstroms is also undoubtedly double, but its duplicity is harder to recognize than that of Hζ.

In Figure 76 the spectra have been exactly aligned with respect to wave length, by means of the comparison lines. It is easily seen that Hβ, Hγ, and the oxygen line at 3727 Ångstroms show practically no displacement between phases 0.32 and 0.82 — near the elongations — whereas the He II line is shifted toward the red at phase 0.82.

From measurements of the Doppler displacements of the outer edges of the He II line at 4686 Ångstroms, Kraft and Greenstein have plotted the radial velocity curve shown in Figure 77. The black dots represent 200-inch observations, open circles those with the

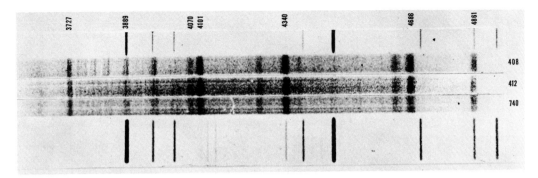

Figure 76. Spectra at phases 0.32, 0.82, and 0.01, aligned according to wave length, to illustrate radial velocity shifts of the helium 4686 line and the changes in appearance of such double lines as Hζ at 3889 Ångstroms.

(*Mount Wilson and Palomar Observatories photograph.*)

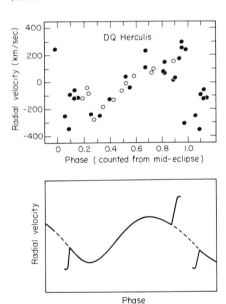

Figure 77. Radial velocities of DQ Herculis, from spectrum lines of ionized helium. In the upper part, dots are 200-inch observations, open circles 100-inch results. The lower part shows the trend: a sinusoidal variation in velocity, with a marked rotational anomaly superimposed.

100-inch. At phase 0.3 the greatest velocity of approach is attained (except for an anomaly discussed in the following paragraphs), and at phase 0.8 the greatest recession, the plotted points forming a sine curve.

This indicates that the nova is moving with a speed of 150 km/sec in a roughly circular path around the center of gravity of the system, at a projected distance of about 250,000 miles. The radial-velocity measurements also indicate that the binary system and the sun are approaching each other at 20 km/sec.

Of particular interest is the anomalous "rotational" distortion of the descending part of the velocity curve. At phase 0.9 the 4686 line is gradually shifting more and more toward the violet end of the spectrum, yet it is suddenly displaced toward longer wave lengths by a velocity of recession of about 300 km/sec. But this effect is very temporary, because immediately after eclipse the line has a velocity of approach of more than 300 km/sec. Thus, the highest and lowest points of the velocity curve occur just before and just after mid-eclipse, respectively.

This phenomenon resembles one previously found in many other spectroscopic binaries, such as U Cephei (Figure 78) and U Sagittae. In the present case the distortion can be explained by the dark companion star eclipsing a gaseous ring that is revolving around

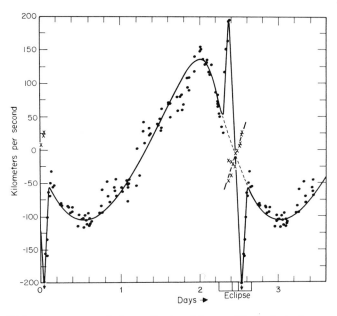

Figure 78. Velocity curve of the eclipsing binary U Cephei showing a rotational distortion before and after mid-eclipse.

the nova component. The unseen star first passes in front of the approaching half of this ring, so only the receding part is seen. After mid-eclipse the reverse is true, with the receding part covered and the approaching half visible.

The ring is revolving at an average rate of 500 km/sec, according to estimates made by Kraft and Greenstein from the total widths of the emission lines produced. The He II 4686 (ionized helium) line and the higher members of the Balmer series originate in the ring, whereas an enormous expanding nebulous envelope is mainly responsible for Hβ, Hγ, and forbidden emission lines, such as OII 3727 (ionized oxygen). According to M. L. Humason, the velocity of expansion of the envelope is about 300 km/sec.

A rapid, pronounced change in the spectrum of DQ Herculis is recorded in Figure 79. Kraft and Greenstein describe this event:

"A further strange phenomenon is seen just before mid-eclipse in the He II emission line, and to a lesser extent in some of the

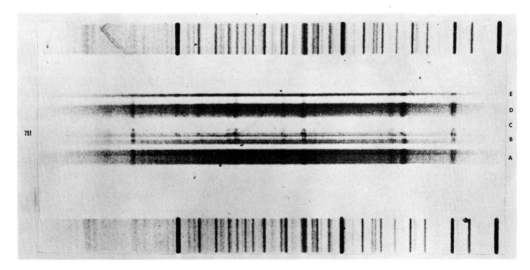

Figure 79. Exposures taken in the order A to E during a primary eclipse of DQ Herculis, at phases 0.89, 0.95, 0.00, 0.10, and 0.15, respectively. A remarkable change in the helium 4686 line (about an inch from the right end) is seen between the first and second parts of exposure B. Within minutes this line's radial velocity altered by 890 km/sec.

(Mount Wilson and Palomar Observatories photograph.)

hydrogen lines, on plate N 751 only. Exposure N 751b is broken into two parts, one centered at phase about 0.94P, and one at phase 0.98P; the latter extends over five minutes. The last two minutes of this exposure show a very sharp break to the shortward (to the violet) when compared with the first part of the exposure — the break amounts to 890 kilometers per second. Little trace of this effect is seen in Hβ, but it is present to some extent in the higher members of the Balmer series."

A tentative explanation of the phenomenon has been proposed by Kraft and is illustrated by the accompanying sketch (Figure 80).

Determining the masses of Nova Herculis and its unseen companion is difficult. If the radial velocity curve of only one component of a spectroscopic binary can be observed — and this is the case for DQ Herculis — it is not possible to determine directly the individual masses of the two stars. However, Table 3, condensed from the work of Kraft and Greenstein, shows the masses corresponding to several arbitrary values for the mass ratio m_1/m_2. Here the subscript 1 refers to the nova, 2 to its companion.

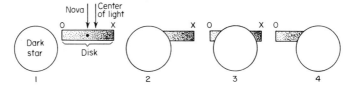

Figure 80. Four stages in a primary eclipse of DQ Herculis, as interpreted in a sketch by R. P. Kraft. He regards the nebulous appendage of the nova as a disk, rather than a ring. Sketch 1 shows the disk just before it begins to be covered by the dark star. In 2, the nova itself is eclipsed, but the brighter trailing end X of the disk is still seen. In 3, the nova is central behind the dark star, and only the ends of the disk send us light. The rapid change in the helium line 4686, shown in Fig. 79, is explained by the covering up of the receding part X of the disk, so only the approaching end O is observable. This last stage is represented in sketch 4, which also corresponds to the lowest point on the light curve, since O is fainter than X.

Table 3. Possible Models of the DQ Herculis System.

Assumed mass-ratio (m_1/m_2)	0.5	1.0	2.0	3.0
Orbital velocity (km/sec)				
Nova	149	149	149	149
Invisible star	75	149	298	447
Mass (solar units)				
Nova	0.075	0.26	1.19	3.18
Invisible star	0.15	0.26	0.60	1.06
Distance in 10^5 centimeters of invisible				
star from center of gravity	2.0	4.0	8.0	12.0

A mass ratio as small as 0.5 seems quite improbable, as the invisible star would then be twice as massive as the nova. Among binary systems in general, the more massive component is usually the more luminous.

An upper limit to the mass ratio can be inferred from the fact that the nova, with its density of 35,000 grams per cubic centimeter, is essentially a white dwarf. We know from the theoretical work of S. Chandrasekhar and M. Schönberg that a white dwarf cannot greatly exceed the sun in mass. Hence, the mass ratio 3, which gives the nova more than 3 solar masses, can be safely excluded. The actual ratio must be near 1 or 2.

The argument can be carried further. If, for example, the nova's mass equaled that of the sun (2×10^{33} grams), we would find for its radius about 2.4×10^9 centimeters — only about $\frac{1}{30}$ the radius of the sun. The eclipse of so small a star by the invisible companion should be exceedingly rapid, yet Walker's light curves actually show a gradual fading toward mid-eclipse, and a gradual brightening afterward. From this, Kraft argues that the total duration of eclipse as observed by Walker corresponds essentially to the eclipse of the gaseous ring. Moreover, the ring contributes most of the light now observed from DQ Herculis, apart from that originating in the large expanding nebulous envelope.

The gaseous ring probably fills most, if not all, of the zero-velocity surface surrounding the nova. These surfaces may be regarded as the boundaries around a binary within which a particle would be permanently retained by the system; a particle outside the

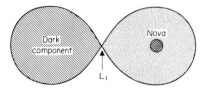

Figure 81. A schematic view of the zero-velocity surfaces surround-ing the two components of DQ Herculis, based on the assumption of equal masses.

zero-velocity surface would tend to escape. In Figure 81 the zero-velocity surfaces have been drawn schematically for equal masses.

Kraft suggests that the invisible star is an M-type dwarf that fills its lobe of the zero-velocity surface, and spills some gas through the Lagrangian point L_1 into the lobe surrounding the nova. This gas becomes the ring revolving around the nova at an observed speed of 500 km/sec. He then demonstrates that this velocity is dynamically consistent with the assumption that the mass ratio is 1, but inconsistent if it is 2. Therefore, according to Table 3, the masses of both the nova and its unseen companion are about 0.26 that of the sun.

Both the radius of the invisible star and that of the outer edge of the gaseous ring turn out to be about 3×10^{10} centimeters. Since the ring is more or less transparent, the observed rotational velocity of 500 km/sec does not refer to its outer edge but to some average distance from the center of the nova, perhaps about 10^{10} centimeters. This is consistent with the requirements of Kepler's third law if the mass of the nova is 0.26.

VI. MAN AND THE UNIVERSE

On October 4, 1957, there occurred an event which, in the course of some 90 minutes — the period of revolution of an artificial satellite around the earth — completely changed the science of astronomy and, with it, the thinking of the entire world. I am referring to the launching of the first successful Russian "sputnik" which is now designated by the symbol 1957 α1.

I have no doubt that because of the Soviet achievement on that date the year 1957 will be remembered forever in the history of astronomical exploration, as the year 1492 is remembered in the history of geographical exploration. There will never again be a lecture on astronomy which does not in some way recognize the enormous achievement of the Russian scientists and engineers in producing the first two artificial satellites of the earth, the first vehicle to land on the moon, and the first photographic camera to circumnavigate the moon, and to telemeter back to us on earth a picture of the "other" side of the moon — the hemisphere that is always oriented away from us and has never been seen by man.

As Professors F. L. Whipple and J. A. Hynek, of the Smithsonian Astrophysical Observatory, stated almost immediately after the launching of Sputnik 1: "In his millennia of looking at the stars, man has never found so exciting a challenge as the year 1957 has suddenly thrust upon him." And a little later Whipple wrote: ". . . the artificial satellites are adding to the knowledge of our planet with every trip around the earth. In the first few weeks of their flight, they told us more about the shape of the earth than 2000 years of observation of our natural satellite, the Moon."

I have mentioned the years 1492 and 1957, and I invite you to make a comparison of the discovery of America by Columbus and the conquest of interplanetary space by the Russian rockets. In the second half of the fifteenth century the leading nation in the field of navigation and geographical exploration was Portugal. Prince Henry the Navigator had assembled at his court the foremost astronomers of his time and had developed astronomical methods of navigation that had enabled Vasco da Gama and many others to undertake long voyages along the western coast of Africa and permitted them, ultimately, to reach the shores of India and China by circumnavigating the continent of Africa.

But when Columbus applied to the King of Portugal, after the death of Prince Henry, for aid in organizing an expedition toward the west across the Atlantic Ocean, he encountered only lack of interest and skepticism. The voyages of Columbus were financed by Queen Isabella of Spain, and from this small investment sprang Spain's great colonial empire, which lasted for several centuries and insured Spain's pre-eminence in the western world until fairly recent times.

Until October 4, 1957, we felt secure in our belief that we were the leading nation in science and engineering. But on that date we suffered a humiliating defeat, and one which future historians will regard as of comparable significance to the defeat which Portugal experienced in 1492 when King John II of Portugal lacked vision and Queen Isabella of Spain possessed it. It is impossible to avoid the conclusion that in 1957 America was lacking in vision and the Soviet Union possessed it.

This does not mean that American scientists were unaware of the importance of space research. I remember that many years before there were any Sputniks in space Professor Whipple told me that many of us would live to see men land on the moon, and at about the same time Professor Lyman Spitzer, of Princeton, urged me and others to work toward the construction of a satellite-borne telescope instead of spending vast sums of money toward the construction of more large ground-based optical telescopes. But their warnings that the conquest of space was imminent went unheeded by the majority of astronomers, their university officials, and by our federal government. We did make some efforts, in a more or less leisurely way, to build space vehicles in connection with the International Geophysical Year because we knew that the exploration of space was coming, but we did not realize that it would be here so soon.

I find it symbolic that the Russians have given the names of *Tsiolkovsky* — their great rocket pioneer — and *Mechta* to two prominent formations on the other side of the moon. The latter word does not, I believe, exist in the English language: it is a combination of the meanings of the words "dream" and "hope," and it represents better than any of their boastful press releases the spirit of science in the Soviet Union today.

As a former Russian, I believe I know why the Soviets have forged ahead of us in the field as astronautics as they have done in several other fields of science and are certain to do in others. In the first place we should remember that Russian science was not invented by the bolsheviks. Ever since the days of Lomonosov — the Russian equivalent of our own Benjamin Franklin — Russian astronomy, mathematics, and physics have been of the highest quality. America's famous astronomer Benjamin Gould once called the Pulkovo Observatory the astronomical capital of the world, and this description was repeated by Simon Newcomb after his visit to Pulkovo in 1870. Yet, despite our traditional admiration for everything Russian (except the czarist and communist governments), until recently no one seemed to reach the logical conclusion that out of the astronomical capital of the world there might come ideas which would amaze the world.

But even more important than the supreme quality of the prerevolutionary education and science in Russia was the tremendous upheaval the country experienced during the 1917 revolution and the civil war. Since the breakup of the Roman Empire no other nation has experienced a similar holocaust. All old and outdated traditions and ideas were swept away, at the cost of millions of lives, leaving the nation free to build a new science on the basis of *Mechta.*

But, as everyone knows, there are many tendencies in the Soviet Union that counteract the free exploitation of all that is meant by the word *Mechta,* and I believe that in the long run the freedom of the democratic society of America will assert itself; we need not fear that our defeat in 1957 will forever prevent us from regaining supremacy in space science; we shall not become a second- or third-rate power, as was the fate of Portugal after its defeat in 1492. To accomplish this we, as scientists, must think hard and work even harder. If the First World War was won by the chemists and the Second by the physicists, the present Cold War is certainly being waged by the astronomers. I shall, therefore, try to explain what astronomy is and what it can contribute to the winning of the Cold War.

Astronomy is a part of physics — all astronomy and not only that part which is commonly called astrophysics. Its purpose is (*a*) to test the operation of known laws of nature under conditions that cannot be realized in a laboratory, and (*b*) to discover new laws of nature that would remain forever unknown because of the very long intervals of time — billions of years — and very great distances — billions of light years — that are required to study them.

Consider the problem of gravitation. The force of universal attraction, discovered by Newton, which pulls two 1-gram masses toward each other is only 6.7×10^{-8} dyne when their centers are separated by 1 centimeter. I suppose this would have remained undiscovered if it were not for the existence of the earth — an astronomical body — whose great mass, 6×10^{27} grams, exerts a force of attraction upon each gram on its surface equal to

$$\frac{6.7 \times 10^{-8} \times 6 \times 10^{27}}{(6.4 \times 10^8)^2} = 1000 \text{ dynes,}$$

and it is this force that keeps us all firmly attached to the surface of the earth and that we must overcome when we shoot a rocket into space.

The force of gravity that holds two massive stars together, causing them to revolve in a binary orbit, is even greater: for Plaskett's famous binary, HD 47129, it amounts to (by Newton's second law of motion) the product of the mass and acceleration. The mass of each component is roughly 50 times that of the sun, or 10^{35} grams, while the orbital acceleration is about 100 cm/sec^2. Hence, $am = 100m = 100 \times 10^{35} = 10^{37}$ dynes. Although we have not yet discovered a close binary system consisting of two white dwarfs, there is no reason why such systems should not be numerous in the Milky Way. The force of gravity acting between the components would be of the order of 10^{41} dynes, the orbital period would be about 30 seconds, and the orbital velocity would be close to that of light. The largest orbital velocity (with respect to the center of gravity of the system) observed to date is about 500 km/sec in the case of the eclipsing binary VV Puppis, whose period, according to G. H. Herbig, is 100 minutes. It is obvious that if we wish to learn more about the properties of gravitation we must turn to astronomical observations. There is, for example, the problem of the change in a binary star orbit in which one or both components lose mass. In the case of a slow, gradual mass loss, as in β Lyrae, the problem has been solved by S. S. Huang and others. There is also some indication of a change in the orbital period of Nova Herculis 1934 during the nova outburst: according to P. Ahnert the period, which is now 4 hours and 39 minutes, was about ½ minute shorter before the outburst. Ahnert estimates that about 10^{-3} of the mass of the nova was exploded into space. It would be even more interesting to study the change in the orbit of a binary if one of the components should undergo a supernova explosion. One can think of even more sophisticated problems connected with the study of gravitation.

Astronomy has contributed much to the study of the outer structure of the atoms. The intense forbidden lines of doubly ionized oxygen and many others in the Orion nebula enabled Ira S. Bowen to explore those electronic transitions which do not occur in a laboratory, because collisions of atoms among themselves and with the walls of their container occur before they have time to fall back into a lower energy state. In interstellar space, where the density of matter is of the order of 1 atom per cubic centimeter and where quanta of radiation are far apart, the atoms remain undisturbed for weeks or years or, in some cases, for millions of years.

Recent advances in nuclear research also had their beginning in astronomy. Astronomers knew, long before there was any thought of an atomic bomb, that the stars are immense atomic reactors and that every gram of solar gas has already produced 10^{17} ergs — vastly more than it could have produced by any process other than a nuclear transformation. They did not know what particular nuclear reaction was operating in the sun; this question was answered almost simultaneously by C. F. von Weizsäcker in Germany and Hans Bethe in the United States. The latter received his inspiration in 1938 at a famous conference in Washington of several physicists and astrophysicists. It was not without reason that our physicists felt confident in advising the government that controlled fusion could be achieved and that the entire earth would not disintegrate in an atomic explosion: they knew from the astronomers that the sun automatically controls its fusion of hydrogen into helium and never exceeds appreciably its output of 2 ergs per second per average gram of its substance.

I shall not enlarge upon the three classical tests of the general theory of relativity proposed by Einstein: the gravitational deflection of the images of stars in the vicinity of the sun during a total solar eclipse, the gravitational red shift of the spectral lines of a small but massive star, and the advance of the orbital perihelion of Mercury. All three tests have been executed many times by different astronomers, and all have given results in accordance with Einstein's prediction. But even today there are a small number of competent astronomers (Erwin Finlay-Freundlich of Scotland is one) who wish to do away with general relativity and are trying to explain the observational results in other ways. The measurement of star positions near the limb of the sun is difficult, and observational errors may be serious. The deflection at the solar limb is 1.75 seconds of arc, but no stars are seen at the limb itself. The amount of displacement falls off rapidly as we go away from the limb. A. A. Mikhailov has pointed out that the radial deflection at the limb of Jupiter amounts to only 0.017 second of arc, but that even this small displacement might be measured with an interferometer when a visual double star is approaching the limb of the planet. If, for example, the separation of the binary is 20″ and Jupiter's radius is also 20″, then, when one component is just grazing the planet's disk and the other is still 20″ away from it, the separation between the components would be 0.″008 less than 20″. Mikhailov and others have also commented on the "focusing" effect that should occur when a distant star is accidently eclipsed by a less-distant one. Although the probability of such an eclipse is exceedingly small in an optical binary, it would be interesting to consider the effect of gravitational bending of light in an ordinary eclipsing variable.

The advance of the perihelion of Mercury is 574 seconds of arc per century, and most of it is caused by gravitational perturbations of its orbit. Only a small remainder of 43 seconds of arc per century remains as the effect of relativity. Conceivably even this small amount could be produced by something resembling the drag in an interplanetary medium.

The red shift of spectral lines amounts to 0.6 km/sec in the case of the sun — an easily observed quantity, but it was found by St. John and others only at the edge of the solar disk. Near the center of the disk the red shift is only 0.24 km/sec. Most of us believe that this discrepancy is caused by the radial streaming-out of gases from the surface of the sun, but we must admit that there is some cause for concern.

In the white dwarfs the red shift would be large. It is proportional to M/R. Since a typical white dwarf has a mass similar to that of the sun, but a radius only about twice that of the earth, or 0.02 that of the sun, the predicted red shift should be about 50 times as large as in the sun, or $50 \times 0.6 = 30$ km/sec. W. S. Adams and J. H. Moore found 20 km/sec for the companion of Sirius, but G. P. Kuiper believes that it may actually be much larger — perhaps 80 or 90 km/sec (because the spectrograms of Sirius B were contaminated by the light of the bright star Sirius A). D. M. Popper found 21 km/sec in the white-dwarf companion of 40 Eridani. Single white dwarfs are not suitable for this test because we cannot separate their own radial motions from the gravitational effect; only in binary systems the motion of the nonwhite-dwarf companion gives us the radial velocity of the entire system.

We should be on the lookout for small gravitational differences in the velocity curves of binary star components of different masses. Since the red shift is proportional to M/R, we might, for example, expect that the velocity curve of a component whose mass and radius are 50 and 5 times, respectively, that of the sun would be shifted by $+0.6 \times 50/5 = 6$ km/sec, while the companion with one-half the mass of the primary and radius twice that of the primary would be shifted by $+0.6 \times 25/10 = 2$ km/sec. The difference of 4 km/sec would be easily observed if the absorption lines of both components are measured. Other tests, by means of space vehicles, have not yet, to my knowledge, been carried out.

Perhaps even more important than the testing of known laws of nature by means of astronomical observations is the discovery of new laws. Best known is the mysterious phenomenon of the expansion of the universe, indicated by the recession of the distant galaxies. I have already mentioned in my third lecture some of the observational results pertaining to this problem. The principal purpose of the 200-inch Palomar reflector was, I believe, to extend the present linear relation between the velocities of recession of the galaxies and their distances. As I have explained, we have encountered great difficulties in the determination of the distances of very remote galaxies, and the Mount Wilson–Palomar astronomers have not yet succeeded in defining this relation with sufficient precision to solve uniquely the cosmological problem of the nature of the universe. The greatest red shift thus far announced by Humason corresponds to a velocity of recession of 60,900 km/sec — about one-fifth of the velocity of light. His work was based upon conventional photographs of distant galaxies of 17th and 18th photovisual magnitudes with a small-dispersion optical spectrograph. But direct photographs of the sky record the images of galaxies down to the 22nd apparent magnitude. Recently, W. A. Baum has used a photoelectric photometer to measure the displacement in wave length of the entire spectral-energy curve of distant galaxies, and has found a red shift in the cluster of galaxies known by its catalogue number 1448 corresponding to 120,000 km/sec — almost one-half of the velocity of light. The photoelectric procedure promises to give even larger red shifts in the next few years.

There is no doubt that astronomy is capable of discovering other fundamental laws of nature. What, for example, happens to all the radiation that is poured out into the universe by the stars? The density of radiation in interstellar space is exceedingly small — about 10^{-12} erg per cubic centimeter. But there are an enormous number of cubic centimeters within our galaxy. The sun alone pours into space 4×10^{33} ergs/second, which corresponds to a mass loss of 4×10^{12} grams per second, or 4 million million tons per second. All 2×10^{11} stars of our galaxy produce about 10^{45} ergs per second, and all galaxies 10^{56} ergs per second. So far as we know now, all of this vast amount of radiation escapes into space at the rate of the velocity of light in all directions and is forever lost from the observable part of the universe. But are we sure that nothing ever happens to a ray of light even if it travels over distances of 10 billion light years? Is it not possible that the quanta of radiation interact with one another, either when they are in flight for billions of years or when they are so densely packed as they are in the neighborhood of a hot star?

Another important law of nature is embodied in the so-called Chandrasekhar limit of the mass of a degenerate white dwarf. This law results from the fact that the equation of state of a degenerate gas, formulated by Fermi and Dirac, requires that the pressure be proportional to the $\frac{5}{3}$ power of the density and be independent of the temperature. In a perfect gas, as in the sun, the pressure is proportional to the density and the temperature. Any star would collapse because of its own gravitation upon itself were it not for the pressure that counteracts the gravitation. If we consider a particular white dwarf, like the companion of Sirius, whose radius is twice that of the earth while its mass is the same as the mass of the sun, the gas pressure in this star exactly balances the internal gravitation. If the mass of Sirius B were doubled, the force of gravity would be increased by a factor of 4. The density would also be increased: if the size of the star remained the same, it would increase by a factor of $2^{5/3}$, or about 3.2. Gravity would exceed the pressure, and the star would readjust itself by becoming smaller: the more massive is a white dwarf, the smaller its radius. When the mass is 1.5 times that of the sun, the radius becomes zero; such a white dwarf cannot exist. Consequently, all white dwarfs have masses less than 1.5 solar masses. If a star had a greater mass in the beginning, it must have discarded the excess before becoming a white dwarf, and we believe that we observe the process of mass-expulsion in nova and super-nova explosions.

Astronomy, aided by physics, has gone a long way toward explaining the origin of the chemical elements and the abundance differences observed in different stars (Figure 82).

These and many other examples illustrate the connection between physics and astronomy. These same questions bring us to the dim frontiers of knowledge where science merges with philosophy. We cannot help asking ourselves: What lies beyond the 10-billion light-year boundary of the observable part of the universe? What is the meaning of the words: the age of the universe; and what happened earlier than 10 billion years ago? The human mind is so conditioned that time appears to us infinite in the past and in the future and that space is Euclidian and not curved. Yet, we also know that the predictions of the theory of relativity have been verified in so many different ways that we cannot disregard the conflict between what we call "common sense" and the abstract mathematical formulation

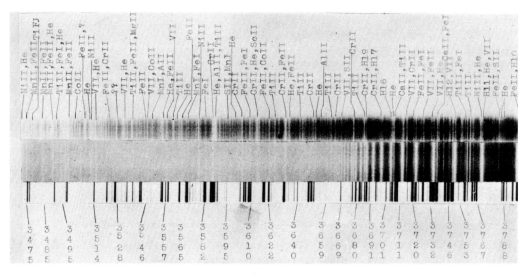

Figure 82. The spectra of two stars: ν Sagittarii, whose atmosphere is deficient in hydrogen (top), and α Cygni, whose atmosphere has a normal, high abundance of hydrogen.

(McDonald Observatory photograph.)

of cosmology. All these questions are important. Most of us have thought about them over and over, especially in our younger years; yet, most of us have found that we could not answer them. I believe that the aging process of a scientist's brain involves an involuntary tendency to shove such questions into his subconscious mind and to deal more and more often with those relatively simple questions that are capable of experimental or observational tests.

I shall, therefore, confine myself here to a quotation from the late Edwin Hubble's George Darwin lecture in 1953 on the expanding universe: "As for the future, it is possible to penetrate still deeper into space — to follow the red shifts still farther back in time, but we are already in the region of diminishing returns; instruments will be increasingly expensive, and progress increasingly slow. The most promising programs for the immediate future accept the observable region as presently defined, hope for only modest extensions in space, but concentrate on increased precision and reliability in the recorded description. The reconnaissance is being followed by an accurate survey, the explorations are pushed towards the next decimal place instead of the next cipher. This procedure promises to reduce the array of possible worlds as surely as did the early rapid inspections of the new territory. And later, perhaps, in a happier generation, when the cost of a battleship can safely be diverted from insurance of survival to the consolations of philosophy, the march outward may be resumed.

"From our home on the earth we look out into the distances and strive to imagine the sort of world into which we are born. Today we have reached far out into space. Our immediate neighborhood we know rather intimately. But with increasing distance our knowledge fades, and fades rapidly, until at the last dim horizon we search among ghostly errors of observations for landmarks that are scarcely more substantial. The search will continue. The urge is older than history. It is not satisfied and it will not be suppressed."

An unfortunate separation of astronomy from physics occurred almost three hundred years ago and has plagued astronomers ever since. Newton was a natural philosopher; but we now recognize the word philosophy only by the granting of Ph.D. degrees. The separation occurred in the seventeenth century, when the practical problem of determining the geographical longitude of a ship at sea became of utmost importance to those national governments which were engaged in transoceanic commerce. The great national observatories at Greenwich and Paris, and many more, were intended primarily for solving navigational problems, not problems of basic science. The astronomers who were connected with the great national observatories, and most astronomers were of this class, were primarily government employees whose salaries were justified by the contribution they made to navigation and geodesy. Even the Pulkovo Observatory in Russia, which was founded in 1839, had the task of training the geodesists of the Russian general staff. The astronomers employed on these tasks were, of course, scientists in the true sense of the word, and they thought of many purely scientific problems. These problems were not spark-plugged by physicists; they arose out of the mass of questions encountered in their more practical duties: accurate star positions, proper motions of stars, and the measurement of time occupied their minds.

Today all of these questions have acquired new significance in physics, but one hundred years ago they were not of great importance to physicists. Even as recently as forty years ago, physics was a relatively unimportant part of an astronomy student's curriculum. The determination of the coordinates of the celestial bodies and the study of celestial mechanics occupied most of his time.

The separation between astronomy and physics had another unfortunate consequence: lulled by two centuries of painstaking accumulation of accurate observations and the very slow growth of our knowledge of the coordinates of the stars, their distances and their motions, we became extraordinarily conservative and unreceptive to new and revolutionary ideas concerning the physical properties of the universe.

I had an occasion to experience the effects of this conservatism ten years ago when the Franklin Institute in Philadelphia was celebrating its one-hundred and twenty-fifth anniversary. A commemorative issue of the *Journal of the Franklin Institute* was intended to "predict future developments in science and technology based upon the considered opinions of today's leading authorities" in order to produce "a composite picture of what can reasonably be expected in the basic physical sciences." I was asked to cover the field of astronomy, and since I am not a good prophet I decided to collect and edit the opinions of about fifty or sixty of the world's leading astronomers. The results were interesting in several ways, but they are even more interesting now than they were ten years ago: only one out of fifty-five leaders of astronomy (Professor M. Minnaert of Utrecht, Holland) mentioned a projectile to the moon; only one other (Dr. Edison Pettit of Mount Wilson Observatory) advocated a revival of interest in the old problem of the canals on Mars. No one mentioned the problem of extraterrestrial life or the possibility of building a space-vehicle telescope. But nearly all predicted "that the boundaries between astronomy, geophysics, physics, electronics, and a number of borderline sciences will be even further broken down" (Professor D. H. Menzel of Harvard University).

If we have been timid in the past, we are now making up for lost time by plunging without concern into the most daring speculations. As Martin Gardner remarked in his book, *Fads and Fallacies in the Name of Science* (Dover Publications, New York, 1957), it is today often difficult to distinguish between science and science fiction. New theories are announced almost every day. They are exhilarating but at the same time dangerous.

Astronomy has had three great revolutions in the past four hundred years: The first was the Copernican revolution that removed the earth from the center of the solar system and placed it 150 million kilometers away from it; the second occurred between 1920 and 1930 when, as a result of the work of H. Shapley and R. J. Trumpler, we realized that the solar system is not at the center of the Milky Way but about 30,000 light years away from it, in a relatively dim spiral arm; the third is occurring now, and, whether we want it or not, we must take part in it. This is the revolution embodied in the question: Are we alone in the universe?

Intuitively we all think of mankind as something unique, something that exists only on earth; and that all the wonders of the universe are intended for *our* benefit and enjoyment. But the vast number of stars that must possess planets, the conclusions of many biologists that life is an inherent property of certain types of complicated molecules or aggregates of molecules, and the uniformity throughout the universe of the chemical elements, the light and heat emitted by solar-type stars, the occurrence of water not only on the earth but on Mars and Venus compel us to revise our thinking.

In all the previous history of science man was able to differentiate clearly between the laws that determine the properties of dead matter and those that involve the recognition of intelligence. Indirect effects of living organisms upon dead matter have, of course, been recognized for some time: we know that the great abundance of oxygen in air is of biogenic origin, as is also the composition of much of the terrestrial solid surface. And we speculate that the exceedingly low abundance of oxygen in the atmosphere of Mars precludes the existence of abundant plant life on that planet today. But I believe that we are beginning rather dimly to perceive that we ourselves are now capable of producing at will various phenomena such as powerful sources of radio waves or brilliant atomic explosions which could be observed from distant planets, outside the solar system; and we must, therefore, revise our thinking and incorporate in our theories possible effects of the free will of other living beings. I am not speaking necessarily of intelligent beings resembling man on distant worlds, but the cumulative action of such phenomena as the migration of animals over the continents of the earth, the seasonal flight of birds, and the appearance and disappearance of many species. On the earth their effects are easily discernible, but they could not now be observed from a distance of many light years. Yet, I believe that this question is one of principle, without regard for what can or cannot now be observed in the galaxy. At the present time the man-made production of powerful beams of radio radiation has certainly produced a change in the physical properties of the earth that could be detected from a distance of 10 or 20 light years. One megawatt of power beamed in a fairly narrow cone would produce a signal that is strong enough to be recorded with existing receivers.

We must, however, distinguish between the probability that stars other than the sun possess families of planets and the probability that there is intelligent life on any of these distant planets. Nearly all astronomers agree that probably many billions of stars in the Milky Way have planets but only a few dozen such stars are closer to us than 20 light years. The probability is also great that a few of these outer planets have some form of life. But the probability that any of them have intelligent life at the present time is vanishingly small. The probability that even if intelligent life now exists outside the solar system, but closer to us than 20 light years, any artificial radio signals are reaching us now is even smaller. But it is not zero, and as Morrison and Cocconi have recently stated in *Nature*, the attempt to record such signals must be made.

Unless we ourselves, through ignorance, smother by man-made radio noise all faint radio waves from outer space, we shall soon be able to record much fainter radio waves than those that reach us now from interstellar space. We shall soon have very large radio telescopes, and even larger antennas are under construction or under consideration. The sensitivity of the electronic receivers which record the intensity of the waves is increasing rapidly. A gain in sensitivity by a factor of 10 would increase the volume of space from which signals could be detected by a factor of about 33. A gain by a factor of 100 — which appears technically possible — would increase the volume by a factor of 1,000. There would then be not one or two dozen stars to examine but tens of thousands, and the probability of finding intelligent beings on a few of their planets would no longer appear hopeless.

But whether or not we shall ever discover artificial radio signals from distant planets, we must face the philosophical consequences of the statement: "We are not alone in the universe." There can be little doubt today that the free will of intelligent beings is not something that exists only on the earth. We must adjust our thinking to this recognition.

Unfortunately, the astronomers have not succeeded in adequately explaining to the public the great importance of protecting on a world-wide basis a number of narrow frequency bands in the radio spectrum for the exploration of the universe. As Professor J. H. Oort of the Netherlands has recently stated, radio astronomy will be a dead science in ten years unless protection is granted.